AF806163

Renewable Energy Communities and the Low Carbon Energy Transition in Europe

Frans H. J. M. Coenen • Thomas Hoppe
Editors

Renewable Energy Communities and the Low Carbon Energy Transition in Europe

palgrave
macmillan

Editors
Frans H. J. M. Coenen
Faculty of Behavioural, Management and Social Sciences
University of Twente
Enschede, The Netherlands

Thomas Hoppe
Faculty of Technology, Policy and Management
Delft University of Technology
Delft, The Netherlands

ISBN 978-3-030-84439-4 ISBN 978-3-030-84440-0 (eBook)
https://doi.org/10.1007/978-3-030-84440-0

© The Editor(s) (if applicable) and The Author(s), under exclusive licence to Springer Nature Switzerland AG 2021
This work is subject to copyright. All rights are solely and exclusively licensed by the Publisher, whether the whole or part of the material is concerned, specifically the rights of translation, reprinting, reuse of illustrations, recitation, broadcasting, reproduction on microfilms or in any other physical way, and transmission or information storage and retrieval, electronic adaptation, computer software, or by similar or dissimilar methodology now known or hereafter developed.
The use of general descriptive names, registered names, trademarks, service marks, etc. in this publication does not imply, even in the absence of a specific statement, that such names are exempt from the relevant protective laws and regulations and therefore free for general use.
The publisher, the authors and the editors are safe to assume that the advice and information in this book are believed to be true and accurate at the date of publication. Neither the publisher nor the authors or the editors give a warranty, expressed or implied, with respect to the material contained herein or for any errors or omissions that may have been made. The publisher remains neutral with regard to jurisdictional claims in published maps and institutional affiliations.

This Palgrave Macmillan imprint is published by the registered company Springer Nature Switzerland AG.
The registered company address is: Gewerbestrasse 11, 6330 Cham, Switzerland

Contents

1 **Introduction** 1
Frans H. J. M. Coenen and Thomas Hoppe

2 **"What Are Energy Communities Under the EU's Clean Energy Package?"** 23
Joshua Roberts

3 **Community Energy in the Eastern Baltic Sea Region: From Standstill to First Steps** 49
Salvatore Ruggiero, Henner Busch, Aljosa Isakovic, and Teis Hansen

4 **Clean Energy Transition in Southeast Europe: The Paradigm of Greece from a Fossil Fuel Mediator to a Community Energy Hub** 75
Nikolaos Ziozas and Theocharis Tsoutsos

5 **The Community Energy Sector in Italy: Historical Perspective and Recent Evolution** 97
Chiara Candelise and Gianluca Ruggieri

6 **Community Energy in Germany: From Technology Pioneers to Professionalisation under Uncertainty** 119
Lars Holstenkamp

7 **Support Structures for Renewable Energy Communities** 153
Beau Warbroek

8 **Energy Communities Promoting Home Energy Savings: Interventions, Theory and Results** 179
Thomas Hoppe and Frans H. J. M. Coenen

9 **Creating an Enabling Policy Framework for Inclusive Energy Communities: A Gender Perspective** 205
Mariëlle Feenstra and Florian Hanke

10 **Housing Communities as Low-Carbon Energy Pioneers; Experiences from The Netherlands** 227
Lidewij Tummers

11 **Conclusion** 257
Frans H. J. M. Coenen and Thomas Hoppe

Index 285

Notes on Contributors

Henner Busch is a Researcher at the Lund University Centre for Sustainability Studies and the Department of Human Geography at Lund University, Sweden. His work focuses on transnational municipal climate networks, community energy development, carbon lock-ins and questions of energy and transitional justice.

Chiara Candelise is Energy Economist and Policy Specialist, at ICEPT, Imperial Centre for Energy Policy and Technology (Imperial College London) and at GREEN, Centre for Research on Geography, Resources, Environment, Energy and Networks (Bocconi University).She is a leading expert on techno-economic and policy analysis of solar technologies, and on community energy and crowdfunding as innovative alternative financing for the energy sector.She also has work experience with both private and public institutions, including the UK Government Department for Environment, Food and Rural Affairs (Defra), and United Nations Development Programme. She founded the first Italian equity crowdfunding platform dedicated to the energy sector.

Frans H. J. M. Coenen is Associate Professor within the Faculty of Behavioural, Management and Social Sciences of the University of Twente. Over the years, his research particular has focused on local and regional sustainable development and capacity development for sustainable development. He takes a special interest in participation in sustainable development in relation to the quality of environmental decision-making. He specialises in environmental policy planning and in public policy evaluation. At this moment, he is involved in several research projects in the field of renewable energy and the local energy transition.

Mariëlle Feenstra is a Senior Consultant and Affiliated Researcher of the University of Twente, the Netherlands. She wrote the first-ever thesis on gender-aware energy policy with case studies from Uganda and South Africa as a University Twente master student in 2002. She worked for fifteen years as a policy advisor for municipalities in the Netherlands specialising in European Affairs. In her PhD research "Gender Just Energy Policy: engendering the energy transition in Europe" (July 2021) she combined her academic interest with her policy experience. Her contribution has been acknowledged both in academia and in practice by invitations to participate in policy formulation activities for the FEMM Committee of the EU Parliament and the European Institute of Gender Equality.

Florian Hanke is a social scientist at the European University Viadrina. He received a PhD scholarship from the Heinrich-Boell Foundation. Focusing on questions of equity and justice in the European energy transition, he investigates energy communities' contributions to a just energy transition. He holds a BA in International Business and Economics and a MA in European Studies.

Teis Hansen is Professor of Innovation and Entrepreneurship at the Department of Food and Resource Economics, University of Copenhagen and Senior Research Scientist at the Department of Technology Management at SINTEF (Trondheim, Norway). His research focuses on sustainability transitions, bioeconomy, regional development, transformative innovation policy, and the foundational economy.

Lars Holstenkamp is co-leader of the junior research group "SteuerBoard Energie" at Leuphana University of Lüneburg's Institute of Sustainability Governance, and board member of the ECOLOG Institute for Social-Ecological Research and Education, Lüneburg. Until 2020, he has been working at the Institute for Banking, Finance and New Venture Management at Leuphana Unviersity. He holds a Dr. rer. Pol. from

Leuphana University. He has studied economics and African languages at the University of Hamburg; and economics, specialising in development economics, at the University of Trier. His research foci include the financing and governance of renewable energies, especially community ownership and sustainable finance.

Thomas Hoppe is Associate Professor within the Faculty of Technology, Policy and Management at Delft University of Technology. His research line is about the governance of energy transition in cities and regions, focusing on low-carbon cities, policy and social innovation. This more specifically includes community energy and co-creation. Dr Hoppe has been involved in multiple European Union funded research projects (i.e., Horizon 2020; Interreg) and has co-edited eight special issues in academic journals. He is chairman of the Platform of *Social Innovation in the Energy Transition*, and is on the Editorial Board of *Energy, Sustainability and Society* (ISSN: 2192-0567).

Aljosa Isakovic is a PhD student at the Department of Geography of Christian-Albrechts-Universität zu Kiel, Germany. His research focuses on regional energy transitions in the Baltic Sea region. By applying concepts from economic geography, the goal of his research is to develop a deeper understanding of the role of intermediaries in regional energy transition processes.

Joshua Roberts is currently Senior Policy Advisor of REScoop.eu (European Federation of Citizen Energy Cooperatives. He led REScoop.eu's advocacy efforts on the Clean Energy for All Europeans legislative package, in particular new provisions that acknowledge and support energy communities as new market actors in the energy transition. A qualified lawyer in California since 2010, Josh has been actively involved in the European energy policy debate since 2012, focusing on issues related to the internal energy market, renewables and other distributed energy resources, infrastructure, and community/citizen participation.

Gianluca Ruggieri is an environmental engineer, and since 2006 he is an Assistant Professor in Applied Thermodynamics and Heat Transfer at the Department of Theoretical and Applied Sciences of Università degli Studi dell'Insubria (Varese, Italy). He took part in several European projects and in countless local initiatives for the promotion of energy sustainability. He is one of the founding members of ènostra, currently acting as vice president. His main research interests include end-use energy efficiency policies and technologies, building materials and building physics, and community energy.

Salvatore Ruggiero is a Research Fellow at the Centre for Consumer Society Research, University of Helsinki, and Postdoctoral Researcher at Aalto University School of Business, Finland. His research concentrates on actors and agency in socio-technical transitions, energy communities, social acceptance of renewable energy technologies, and business model innovation for energy flexibility.

Theocharis Tsoutsos is a Professor in the Chemical & Environmental Engineering School, Technical University of Crete, Director of Graduate Programme "Environmental Engineering" (2014–2018), as well as Founder and Director of the Renewable and Sustainable Energy Lab (2005–). He has taught in universities such as the Open University (UK), Kingston, Herriot-Watt, West Attica, and the Greek Open University. Theocharis has working in Sustainable Energy and Mobility for 30+ years and has coordinated 50+ European projects. He was Head of the Development Department of the Centre for Renewable Energy Sources and Saving—CRES (1992–2005).Theocharis is the author of 400+ scientific publications in international journals and proceedings, and has achieved 5000+ citations (Scholar) with h-factor:30(Scopus) and 36(Scholar). He is currently an editorial board member of Sustainable Cities and Society (Elsevier); Energies(mdpi); AIMS Energy; Current Sustainable/Renewable Energy Reports(Springer-Nature).

Lidewij Tummers commutes between Rotterdam and Hannover. In 2000, she founded the office Tussen Ruimte (Intermediate Space) in Rotterdam (NL), and has since specialised as a technical advisor for self-organised housing, gender+diversity in spatial planning, and climate-neutral building. Tussen Ruimte advises building groups in building and technology decisions and planning departments in European cities and regions on integrating self-organised housing communities in sustainable urban development.Since 2006, Tummers has been teaching at various European architecture and urban planning faculties. Her dissertation

(TU Delft, 2017) applied a feminist approach to investigate the transformative potential of self-organised housing in the European context, by the example of energy systems.

Beau Warbroek is Postdoctoral Researcher within the Department of Civil Engineering at the University of Twente. He holds a PhD in Policy Studies from the Department of Governance and Technology for Sustainability (CSTM) at the same university. Dr Warbroek's research line is about the citizen-based energy transition, in particular focusing on low-carbon energy initiatives. He has multiple publications in international peer-reviewed journals (including *Energy Research & Social Science*) and has presented his work at several international conferences, such as the International Conference on Public Policy and the Royal Geographical Society (with the Institute of British Geographers) International Conference.

Nikolaos Ziozas is a Research Assistant at the Renewable and Sustainable Energy Systems Laboratory (ReSEL), Technical University of Crete. Nikolaos holds a Diploma in Mechanical Engineering from the Aristotle University of Thessaloniki, and specialises in building energy design. His main research interest is the interdisciplinary examination of the ongoing community energy revolution as well as the critical role of civil society within it. His main research projects draw upon theories and methodologies from socio-technical energy transitions and the energy policy field, focusing on the Multi Level Perspective and Strategic Niche Management frameworks. His MSc thesis, entitled "The Role of Civil Society in Community Energy niche: A case study of Greece", was the first research on the emergence of community energy in Greece from both an empirical and socio-technical approach.

List of Figures

Fig. 3.1	Strength of key contextual conditions in the Baltic Sea Region	56
Fig. 4.1	Phases of the development of shared technological knowledge. (*Source*: Geels & Deuten, 2006)	79
Fig. 4.2	Map of the examined REScoops in the Greek area	81
Fig. 6.1	Types of community energy in Germany across time	128
Fig. 8.1	Theoretical framework on interventions regarding home energy savings. (Adapted from: Ajzen (1991); Coenen and Hoppe (2018))	185
Fig. 8.2	Venn diagram of energy community membership (here applied to REScoops), discerning between general membership, engagement in REScoop activities and participation in specific dedicated REScoop actions. (From: Coenen & Hoppe, 2018)	188
Fig. 8.3	Classification of measures implemented by energy communities to persuade households to save energy	191
Fig. 8.4	Overview of members participating in dedicated energy savings actions implemented by energy cooperative LochemEnergie (in relative frequency of total membership)	195
Fig. 10.1	TERBREGSE.NL (Drawing by Hulshof Architekten, 2000, inscriptions by author)	237

List of Tables

Table 2.1	Energy Community definitions in the CEP	27
Table 3.1	List of CE cases, countries, technology used, and main actors involved	55
Table 3.2	Strength of historical and current energy policies in the Baltic Sea Region	58
Table 3.3	Summary of findings from the case studies	64
Table 4.1	Descriptive information of the three examined Greek REScoops	83
Table 6.1	Major developments in the German legal framework related to renewable energies and energy communities	122
Table 6.2	Comparison of definitions of community energy/energy communities in Germany	135
Table 6.3	Selected community energy-related activities in the German federal states	139
Table 7.1	Overview of intermediary strategies	163
Table 7.2	Overview of intermediary roles and activities	169
Table 8.1	Overview of specific dedicated measures used by energy communities	192
Table 9.1	Persistent rights failures in the energy system through a gender lens	211
Table 10.1	Varieties of Dutch resident-led housing (author, August 2014) Updated 2021	229

Table 10.2	Significant environmental policy periods and pilot projects experimenting with sustainable building	232
Table 10.3	Overview of energy-features of discussed projects	234

1

Introduction

Frans H. J. M. Coenen and Thomas Hoppe

1.1 Trends and Observations in European Renewable Energy Communities

The use of renewable energy helps to mitigating climate change. Besides generating energy that avoids emitting greenhouse gases and does not cause air pollution,[1] it has a much wider impact on society. The economic

[1] Next to burning wood locally, which is evidenced to seriously pollute the air and contribute to problems related to nitrogen emissions. Moreover, burning wood can cause a nuisance to local residents in the form of odour nuisance, health problems and soot deposits. URL: https://www.rivm.nl/toolkit-houtrook-en-gezondheid/vraag-en-antwoord-houtrook

F. H. J. M. Coenen (✉)
Faculty of Behavioural, Management and Social Sciences, University of Twente, Enschede, The Netherlands
e-mail: f.h.j.m.coenen@utwente.nl

T. Hoppe
Faculty of Technology, Policy and Management, Delft University of Technology, Delft, The Netherlands
e-mail: t.hoppe@tudelft.nl

© The Author(s), under exclusive license to Springer Nature Switzerland AG 2021

F. H. J. M. Coenen, T. Hoppe (eds.), *Renewable Energy Communities and the Low Carbon Energy Transition in Europe*, https://doi.org/10.1007/978-3-030-84440-0_1

impact results from diversification of the energy supply, in reducing dependence on imported fuels, and in leading to new economic activities that create jobs in manufacturing, installation, and more. Furthermore, there are positive social aspects such as stronger community solidarity and spirit, because many renewable energy activities take place within local communities. Forms of renewable energy, like solar power and wind energy, do not always need a centralized energy systems approach, and can be scaled down and used to serve the needs of local communities. Moreover, this does not need to depend on the energy production and supply from large energy supplying companies. Local communities can reap the benefits from distributed generation, and technically speaking have the possibility to manage and operate decentralized energy systems themselves. In this sense, renewable energy could be seen as a way to empower local communities. Because of this possibility of decentralized, renewable energy generation organized within local communities, it is often associated with the concept of community energy. For groups of citizens and communities that collectively organize decentralized energy actions (with application in the domains of renewable energy, energy efficiency, and energy service provision) the concept 'Community Energy' (CE) is increasingly used among scholars in academic debates as well as in policy practice. However, CE is often used and widely interpreted by an increasing number of scholars. As a consequence, there are numerous definitions and meanings abound, which are caused by a wide variety of activities and projects that each contribute to the plurality of the concept (Brummer, 2018). Nonetheless, an often-used definition describes CE as a wide range of citizen initiatives that are concerned with greening (parts of) energy systems, usually but not only limited to the local or regional scale (Bomberg & McEwen, 2012).

Through the "Clean energy for all Europeans" package (CEP), the EU has introduced the concept of 'energy communities' in its legislation, notably as: (a) 'citizen energy communities' (Article 2 of the Electricity Directive) and (b) 'renewable energy communities' (RECs) (Article 2 of the Renewable Energy Directive). In this edited volume, we address REC initiatives in different continental European countries in the light of the revised EU Renewables Directives. The aim of this introductory chapter is not to provide a systematic literature review on renewable energy community initiatives, but to position the goal and questions of this volume in relevant academic debates surrounding the growing academic research domain of CE, contributing to "a bulwark of empirical cases, theoretical reflections and methodological tools" (Creamer et al., 2019, p. 1). Neither does this chapter aim to provide an overview of definitions and characteristics of these citizen initiatives that go by very different names in the academic literature and policy practice, like citizen-led renewable energy initiatives, local renewable energy organizations (LREOs), local low-carbon energy initiatives (LLCEIs) or (even) renewable energy supplying cooperatives (RESCoops). In this volume we use the Renewable Energy Directive concept of 'renewable energy communities' (RECs) for all citizen-led renewable energy initiatives. In Sect. 1.2 of this chapter, the renewable energy communities' definition and the characteristics of the EU directives and policy are discussed. The defining characteristics of the EU concepts of renewable CE initiatives and the principles behind the EU policy towards citizens and energy communities, will be confronted with four general observations.

Community initiatives seeking to produce, distribute and consume energy locally are not a new phenomenon. For example, rural electrification cooperatives have existed in a number of remote areas within the former German Reich in the first half of the twentieth century (Yildiz et al., 2015) and in the United States (Yadoo & Cruickshank, 2010). Since the 1980s, community-owned renewable energy projects have increased in number considerably, in particular wind energy projects. Although the first known wind turbine producing electricity was already erected in 1887, it was only at the beginning of the 1980s that the first wind energy initiatives in continental Europe (e.g., Netherlands, Demark) started. At first, the initiatives involved university researchers, and were soon followed by initiatives of environmental groups and citizens starting up so-called wind cooperatives (Hufen & Koppenjan, 2015). This is considered the start of the CE movement of active citizens in the way it is currently acknowledged.

More recently, renewable energy community (REC) initiatives have emerged. This has been extensively documented and discussed in the academic literature (Arentsen & Bellekom, 2014; Hoffman & High-Pippert, 2010; Hufen & Koppenjan, 2015; Oteman et al., 2014, 2017; Seyfang

& Haxeltine, 2012, Seyfang et al., 2014; Soeiro & Dias, 2019). What gets much attention here is that these energy initiatives not only consist of citizens who jointly seek to produce and consume sustainable energy locally, but that these citizens and initiatives typically reject the use of centralized-produced fossil and nuclear energy, and embrace local egalitarian values (Hoppe et al., 2015; Oteman et al., 2014). Therefore, these initiatives are not only instrumental in contributing to sustainable energy transition goals in producing renewable energy, lowering carbon emissions and involving citizens. RECs also have more normative functions for citizens and their energy communities than merely producing renewable energy collectively. There are specific reasons for their existence, linked with objections against current practices in current energy markets, and more generally the fossil fuel and nuclear centralized energy system model. In the European Union, there is great diversity in the number of REC initiatives per country. The numbers differ according to the definitions used. In 2021 the European federation for renewable energy cooperatives (Rescoop.eu or Rescoop Europe) had a growing network of over 1900 members, i.e., local energy cooperatives (Rescoop.eu, 2021). Rescoop.eu federation members have signed the Rescoop.eu Charter, based on the cooperative definition (see Sect. 1.3). Given the differences in definitions of RECs it is arbitrary to give numbers of initiatives per country. Overviews have been constructed based on national official registries of energy cooperatives (Wierling et al., 2018) or through a combination of official registries and literatures (Hewitt et al., 2019). Membership registrations and registries might seem a more objective source, but still the question arises what REC actually means and how it is perceived. For instance, are Polish energy 'clusters' actually community initiatives, given the fact that they are often only established by either local authorities or private sector commercial initiatives (see Chap. 3)?

Another, maybe even more arbitrary perspective is to think in terms of clusters or groups of countries as front-runners, late starters or as a middle group. This in analogy to how the implementation of LA21 as Chapter 28 of the Rio Agenda 21 in Europe was categorized in the 1990s (Lafferty & Coenen, 2001). Positions in the different groups might change over time. In 2014, the European front-runners in CE were clearly Germany, Denmark and Austria (Oteman et al., 2014). On the basis of the overview articles and additional sources countries like Italy (17) (Candelise & Ruggieri, 2020; Candelise and Ruggieri Chap. 5 in this volume) and Spain (19) (Unión Renovables Coop, 2021) and Poland (Hewitt et al., 2019) with around 20–30 initiatives in the late starters group can be identified. Next, there is the middle group of countries with more than one hundred initiatives like Sweden with 112 (Kooij et al., 2018), the Netherlands with 623 initiatives (Schwencke, 2021) and the UK with over 300 in England and Wales (Community Energy England, 2020) and particular Scotland leading with 416 initiatives (Community Energy Scotland, 2021). Front-runners would be countries with around 1000 initiatives like Denmark (1109) and Germany (965); entries in the database by Wierling et al., (2018). Comparatively, the Netherlands has been catching up with the front-running countries. Belgium and France are difficult to position in between other countries, comparatively, because the CE uses another organizational model. By 2021, Belgium (Hitchins et al., 2020; Bauwens, 2016) had only around thirty initiatives and France just over seventy (Hewitt et al., 2019). However, in both countries the energy cooperatives have high membership rates (i.e., Ecopower, Belgium: over 50,000; Enercoop, France: 42,000) (Hoppe & Coenen, 2016).

In the Netherlands, the 623 local energy initiatives comprise nearly 100,000 participants (Schwencke, 2021). Therefore, the number of initiatives per country alone does not say everything about the significance of community energy. In 2021, Rescoop.eu, as a federation representing the community energy movement, accounted for over 1900 cooperatives across Europe, which included membership of over 1,250,000 citizens (rescoop.eu/network, 2021). Given the specific membership requirements REScoop.eu uses, some countries are over-represented. Another way to look at significance would be the type of renewable energy of the initiatives and the total share of the renewable energy they produce. For instance, in the Netherlands the initiatives generate 1.6 per cent of the total amount of solar energy and 5.7 per cent of wind energy (Schwencke, 2021). Another perspective is the growth of the significance. Again, the Netherlands as an example, shows a doubling within four years with an increase from 332 and 50,000 respectively in 2016 (Schwencke, 2017) to 623 and nearly 100,000 participants in 2020 (Schwencke, 2021). Finally, the growing importance of RECs—besides increased numbers of cooperatives and their membership numbers—can also be indicated by increased ownership numbers of renewable energy plants. In Denmark a large share of the wind projects is actually owned by the public (Gorroño-Albizu et al., 2019). In the Netherlands the Danish community ownership model of wind parks was recently adopted following effective lobbying by the renewable energy community branch organization. 50 per cent local ownership was agreed upon in the national 'Climate Agreement' (the nation's metagovernance approach to climate mitigation) to foster local ownership of solar and wind energy parks (Schwencke, 2021).

Between 2010 and 2020 the growth of renewable energy communities has gone hand in hand with increasing academic interest, research and publications from different academic perspectives and research agendas. Typical research themes include the difference in the organizational model of renewable energy communities (Creamer et al., 2019), the motives of their members (Dóci & Vasileiadou, 2015; Bauwens, 2016) or, barriers for starting initiatives (Boon & Dieperink, 2014). Furthermore, there is the supportive role of local government vis-à-vis renewable energy communities (Hoppe et al., 2015), the support structure of initiatives (Warbroek et al., 2018) and specific economic or fiscal policy instruments to support renewable energy communities (Kooij et al., 2018). In addition, there is the relation with social innovation (Hewitt et al., 2019), with social capital (Bauwens & Defourny, 2017), social entrepreneurship (Becker et al., 2017; van der Horst, 2008), and the importance of policy support to community energy initiatives for the energy transition (Wierling et al., 2018). In particular, this last aspect gives a direct link with EU directives, in which renewable energy community initiatives form the core of energy transition policy. In the next sections the EU policy initiatives and new directives will be introduced and discussed. There is also a growing academic interest in the CEP and the EU directives; their characteristics, how they were established, implications and implementation aspects (Roberts, 2020; Cobutt, 2021). This volume contributes to the body of literature on CE.

1.2 Four Observations

This volume starts from four observations of developments in community energy. The first observation is that community energy is getting a more instrumental function in the energy transition. The EU is now taking a leading role in community-owned energy projects since this has emerged as common practice in some EU Member States, but certainly not in all countries. The future scenario that the EU would like to see enrolled and implemented across all of its Member States is that community energy and energy active citizens are key for the energy transition in all EU Member States, regardless if they are front-runners or relative late starters in prioritizing community energy. This is not about sheer numbers of initiatives or the share of community energy in renewable energy production, but about the potential of community energy for the EU energy transition. This view on the importance of energy communities and consumer (co-)ownership in renewable energy production as essential for the overall success of the energy transition gives community energy a more instrumental function. This does not draw the attention away from the fact that there are also many normative societal reasons to stimulate community energy, like views on energy democracy, energy and climate awareness, active citizenship or community empowerment (Van Veelen & van der Horst, 2018).

A growing number of scholars argue that local energy communities are increasingly important players in sustainable energy generation and energy saving efforts (Bauwens, 2016; Bauwens et al., 2016; Dóci et al., 2015; Seyfang et al., 2014; Soeiro & Dias, 2019). Moreover, the EU policy initiative came about thanks to front-running CE initiatives lobbying themselves (Roberts, 2020), using instrumental and normative arguments to address importance of community energy.

RECs are not only instrumental in terms as being merely a means to achieve sustainable energy transition goals (e.g., contributing to lowering of carbon emissions). They also have a more normative function for the citizens within these initiatives. These normative reasons are linked with objections against the ways existing energy markets operate, and in particularly to large-scale (fossil fuel) energy systems. Rather than profit-seeking firms, RECs are embedded in local structures that have certain collective values, and their resources entail voluntary input, grant funding and reciprocal relations. Nor do they seek profit maximization (Warbroek & Hoppe, 2017). There are also specific anti-elitist reasons for the start-up of these initiatives, like dissatisfaction with the government and a dissent with big players in the energy industry (Bauwens, 2016; Hoppe et al., 2019). In addition, certain societal or altruistic motives apply like ecological motives or the liveability in the region or seeing the community energy movement as part of regional social capital (Bauwens & Defourny, 2017). Other motives are instrumental for the participants themselves, like wanting to be self-sufficient, economic motives or seeking civic and social gratification.

A second observation is that there is a clear difference in the starting situation between front-runners and late starters, both between countries and between initiatives themselves. Late starters do not have to "reinvent the wheel": there is learning and access to information. However, whether countries can be seen as relative starters or front-runners, they all face the same directives. Bauwens, Gotchev and Holstenkamp (2016) discern several important factors which are intended to clarify the disparities between EU countries, such as formal institutional rules, the support mechanisms for renewable energy (RE) and spatial planning, attitudes toward the cooperative model and the cultures of local energy activism. These factors that influenced the emergence of CE initiatives in countries are still important for the differences between the European countries, but get another meaning in the light of the new EU policy. If all EU countries face the same directives, what would the new EU policy mean for countries that are late starters in the CE movement? Here, a growing attention in for example Southern European countries for community energy is observed (Rescoop.eu, 2021), but the EU policy initiative is not the only explanatory variable for this growing attention in these late-starting countries. On the other hand, front-runners who are confronted with the new directives may face the "Law of the inhibitory lead", where having a head start might be considered a handicap. And how does this play out for renewable energy community initiatives' members that started from idealistic motives, and are now joined by new members, i.e., people with different motives (Bauwens, 2016). A crucial factor that determines the difference in starting situations of REC initiatives is the role of government (Oteman et al., 2014; Hoppe et al., 2015; Warbroek & Hoppe, 2017) and its promotion, empowerment (or hindrance) of CE. Government on different scales can enable initiatives by facilitating, coordinating and encouraging action through financial incentives, public–private/voluntary partnerships, shaping policy goals in partnership and providing infrastructure.

A third observation is the broadening of the goals of the CE movement and the explicit recognition that CE is not just about jointly producing renewable energy. This is particularly important in light of raising awareness and attracting public support for low-carbon energy and energy saving (Warbroek & Hoppe, 2017). REScoop.eu, defines renewable energy cooperatives as, "groups of citizens who organize themselves to collectively take action to foster the use of renewable energy and increase energy efficiency standards" (REScoop.eu, 2016). In their quest to increase energy efficiency standards they engage in activities like persuading their members to conserve energy (e.g., Coenen et al., 2017; Hoppe et al., 2016; Oteman et al., 2014; Van der Schoor & Scholtens, 2015; Hoppe et al., 2019). Apart from energy demand, CE initiatives are also active in other activities like promoting smart grid experimentation, or starting EV car sharing business models (see Chap. 8 of this volume).

A fourth observation concerns a mismatch between academic attention and the significance of RECs in certain countries. A high share of scholarly literature on community energy and local energy initiatives focused on practices in the United Kingdom (Van der Schoor & Scholtens, 2021), which is home to comparatively few energy cooperatives—yet still has over 300 organizations (Community Energy England, 2020, p. 3)—when compared to other European countries. Countries in continental Europe with relatively high numbers of renewable energy communities like Germany, the Netherlands and particular Denmark received relatively less attention. In addition, a lack of literature is observed on some of the late-starting countries. Therefore, continental Europe serves as the locus of the research presented in the present volume. Next to attention to front-running countries specific attention is also given to some of the late starters.

These four observations form the structural basis of this volume. This entails: (1) a focus on a more instrumental function of CE in light of the EU sustainable energy transition; (2) differences between front-runners and late starters in a starting position who are faced with the revised EU Directives; (3) the broadening of goals and activities of RECs, and; (4) the mismatch between scholarly attention and significance of the CE movement in countries, to formulate four sub-questions. Before presenting the main research question, first EU policy initiatives are addressed and then EU policy initiatives are confronted with the four central observations that are central to this volume.

1.3 Towards Community Energy Projects as Common EU Practice

On 14 June 2018 the European Parliament and the Council finally reached a political agreement on rules addressing how the EU will roll out renewable energy over the next decade. The EU decided on a binding objective of increasing renewables by 32 per cent to be achieved in 2030, with the possibility to review the target in 2023 in order to revise it upward. This presents an unprecedented development as the revised EU Renewables Directive (2009/28/EC) provides explicit and well-defined roles for citizens and communities. The EU discerns renewable energy communities (RECs) from citizen energy communities (CECs).

Article 2 of the Renewables Directive provides a definition of renewable energy communities. 'Renewable energy community' means a legal entity:

- which, in accordance with the applicable national law, is based on open and voluntary participation, is autonomous, and is effectively controlled by shareholders or members that are located in the proximity of the renewable energy projects owned and developed by that legal entity;

- has for its primary purpose to provide environmental, economic or social community benefits to its members or shareholders or to the local areas where it operates rather than to generate financial profits; and
- may engage in electricity generation, distribution and supply, consumption, aggregation, storage or energy efficiency services, generation of renewable electricity, charging services for electric vehicles or provide other energy services to its shareholders or members."

The definition of 'Citizen energy communities' (Article 2 of the Electricity Directive) uses a similar definition. The second and third bulleted issues are the same in as the Renewable Energy Directive. The first issue holds that the autonomy principle is absent when compared to the definition of RECs.

By making this agreement, the EU has created a landmark of citizen empowerment in the face of low carbon energy transition. More specifically the agreement enables and empowers renewable energy cooperatives (REScoops), which are advocating low carbon community energy, prosumerism and energy democracy. The agreement can be seen as a milestone for these civil society organization who have been fighting for more and legalized empowerment for decades. This was coined by the President of REScoop.eu, Mr. Dirk Vansintjan who stated, "This is a remarkable day for European energy citizens. Up until yesterday they had no recognition in Europe's energy policy. Now, they have a set of tools to empower themselves and their communities so that they can prosper in the energy transition." (REScoop.eu, 2018).

The next question would be: what is changing for renewable energy communities with the revised Directives? Among others, it contains definitions of 'renewable energy communities' and 'self-consumption'. Next to setting strong definitions the Directive provides rights and a basis for EU Member States to develop national legislation and regulatory frameworks to acknowledge, govern and support renewable energy communities and self-consumers (i.e., 'prosumers'). This includes taking into account RECs in national renewable energy support schemes, improved collaboration between the latter and local authorities, and supporting and strengthening the role of RECs in helping (socio-economic) vulnerable customers and alleviation of poverty. In addition, the revised

Renewables Directive also contains rules to exempt 'prosuming' households who do not feed their excess supply of electricity into the grid, from (conventional) grid charges (ibid.).

Against the background of the general objectives of keeping the EU a global leader in renewables and, more broadly, helping the EU to meet its emissions reduction commitments under the Paris Agreement (COP21), a new and extended role of citizens and renewable energy communities play an instrumental role in achieving these objectives. In relation to our first observation of the shift from a more normative perspective of the role of citizens and renewable energy communities in the energy transition to a more instrumental role, elements of this instrumental function are that energy communities:

- will help to pave the way for a clean energy transition, while moving citizens to the fore because they organize collective and citizen-driven energy actions;
- will contribute to increase of public acceptance of renewable energy projects;
- will make it easier to attract private investment in the clean energy transition;
- have the potential to provide direct benefits to citizens by advancing energy efficiency and lowering their electricity bills;
- can help in providing flexibility to the electricity system through demand–response and storage by supporting citizen participation.

Formulated in this way energy, communities are instrumental for the energy transition. However, in the definition of renewable energy communities one also recognizes more normative elements based on open and voluntary participation, autonomously and effectively controlled by local shareholders or members and providing environmental, economic or social community benefits to its members or shareholders or to the local areas where it operates rather than to generate financial profits.

Here, one clearly recognizes the ideas of the cooperative movement and the definition of the cooperative organization. The International Co-operative Alliance (ICA) defines a cooperative as, "an autonomous association of persons united voluntarily to meet their common economic, social, and cultural needs and aspirations through a jointly owned and democratically-controlled enterprise" (ICA, 2017). Seven cooperative principles are used as guidelines by which cooperatives put their values into practice: (i) voluntary and open membership; (ii) democratic member control; (iii) economic participation of members; (iv) autonomy and independence; (v) education, training and information; (vi) cooperation among cooperatives, and (vii) concern for the community (ibid.). According to the ICA, cooperatives are based on the values of self-help, self-responsibility, democracy, equality, equity and solidarity. In the tradition of their founders, cooperative members believe in the ethical values of honesty, openness, social responsibility and caring for others. These principles also apply to renewable energy community organizations. For example, all federation members of Rescoop.eu have to sign the Rescoop.eu Charter, which is based on the cooperative definition, related to its core values and the seven principles of the ICA. The Rescoop.eu Charter added its own ecological principles to the ICA principles and values, that stress the need to protect the environment and social, and ethical principles that highlight the importance to support local economy and prosumers, to ensure financial transparency, to restrict the exclusive pursuit of financial profit, to encourage fair trade and to get access to common goods (Rescoop.eu, 2013).

When reflecting on our second observation there is a clear difference in the starting situation between front-running and late-starting countries. The Clean Energy Package and in particular the revised EU Renewables Directive (2009/28/EC) also asks for some steps from national governments to support the development of community energy and strengthens the role of the citizens in the energy transition. Under the Regulation on the Governance of the Energy Union and Climate Action (EU) 2018/1999, EU countries were required to draft national energy and climate plans (NECPs) for 2021–2030, outlining how they will meet the new 2030 targets for renewable energy and for energy efficiency. Most of the other elements in the new directive had to be transposed into national law by Member States by 30 June 2021, when the original renewables directive would be repealed.

When regarding the third observation—i.e., observing a broadening of the goals of the community energy movement between 2011 and

2020—this is clearly reflected in the third sub-paragraph of Article 2 of the RED II Directive that addressed goals other than renewable energy production-related activities like energy efficiency services and charging services for electric vehicles. This third paragraph of Article 2 means an explicit recognition that community energy is not just about jointly producing renewable energy, it also provides the legal legitimacy to renewable energy communities for broadening their activities. The transition towards a low-carbon economy is not just about a growing role for RE sources, but also about greater energy efficiency and the electrification of transport and other sectors. For prosumers, next to producing electricity for self-consumption and feeding excess energy into the grid, more attention to energy efficiency stresses the role of consumers as different in managing their demand actively. Both prosumers and renewable energy communities can play a role in influencing their memberships (Hoppe et al., 2019; Horstink et al., 2021).

1.4 Renewable Energy Community Initiatives in the Light of the Revised Renewables Directive

The central research question in this volume is: *How did renewable energy communities in continental Europe evolve, and what can be learned from these experiences for the realization and implementation of the Clean Energy Package?* Our four sub-questions relate to the previous four observations in Sect. 1.1 and elaboration of these observations based on the EU directives.

Based on our first observation the first sub-question is: *What does the general trend of a greater focus on a more instrumental function of community energy for the EU energy transition mean for the other more normative functions of renewable energy cooperatives?* For instance, does it influence the value and principles of their cooperative identity? Does a growing significance of the CE movement replace their idealistic motives with pragmatism?

Based on the second observation that there are clear differences between front-running and late-starting countries in their starting position but they are faced with the same directive, the second sub-question is: *What does the new EU policy mean for countries that are late starters or front-runners in the community energy movement?* Attention is paid to how renewable energy communities faired in the period in which previous EU Directives were issued and implemented, through transposing the EU directives in national legislation by EU Member States. This is done by presenting cases from different EU countries. Finally, the ways in which the revised Renewable Energy Directive would enable RECs in selected EU countries that were presented previously, and in which ways this contributes to these initiatives attaining their own goals, and to low carbon energy transition in Europe in general are explored. Besides mentioning support structures, the challenges that the implementation process of the revised Renewables Directive to national legislation might encounter, which might obstruct implementation in support of renewable energy communities are highlighted.

The third observation about the broadening of goals and activities of renewable energy communities over the years brings us to the third sub-question: *What does the broadening of their goals and activities mean for renewable energy communities and what role does the revised Directive play?* In the discussion about paragraph 3 of Article 2 of the Revised Renewable Energy Directive we suggested that this third paragraph means an explicit recognition that community energy is not just about jointly producing renewable energy and gives a legitimacy for renewable energy communities to broaden their activities. However, is this actual the case in the different countries? To give citizens and RECs a fair level playing field, electricity markets, regulations and technical infrastructures need to be adapted. Markets, regulations and technical infrastructure at this moment are still highly intertangled with incumbent regime players like large utilities and large-sized energy companies that dominate domestic and regional energy markets. It is not just the redesign of markets that is needed to encourage citizen to invest in RE. Additional, new regulations are required if the RECs' goals are to be broadened; for instance national regulation on charging services for RECs and their members, the (maintained) use of feed-in tariffs, or subsidies or tax measures that support renewable energy communities developing and exploiting collective RE projects. Other activities also request new and additional business models that RECs can use. Both in prosumerism and energy savings, RECs can play a role, while at the same time keeping costs for their members under control (Hoppe et al., 2019). Important mechanisms are trust in their own RECs and social norm-setting among their members (Coenen et al., 2017).

Finally, based on the fourth observation that there is a mismatch between academic attention and significance of the community energy movement in countries, a fourth sub-question is formulated: *Whether more explicit attention to continental Europe and particular late-starting countries will bring new insights?* For instance, are the barriers for successful RECs the same for countries that received less attention thus far in the academic literature? In this volume, attention is paid to the history and development of the REC movement in continental Europe: their vision and beliefs, ways of operating and in particular, the challenges and barriers they encounter. The latter pertains to RECs' involvement in energy markets where they are confronted with existing regimes, incumbent actors, structures and regulatory frameworks. With regard to countries with a low degree of REC action, or countries where the development of RECs is relatively slow like in Southern-European countries. Because these countries are less researched in the academic literature, do they present a different picture? Furthermore, possible explanations for differences between front-runners and late-starting counties are compared with the previously formulated explanations on RECs disseminating across EU Members State countries (Bauwens et al., 2016).

1.5 Contour of the Volume

The contributors to this volume are all researchers in the field of community energy. The chapters they provide in a loose sense represent over six EU countries (Germany, Greece, Italy, Finland, Poland and some of the Baltic States, and Netherlands). Seven countries are part of the country case study chapters. In total, the book contains information from eleven countries in continental Europe.

Chapter 2 by Joshua Roberts discusses what energy communities are in the Clean Energy Package and what changes it brings to the world of EC. The next four chapters address specific countries or groups of countries. Chapter 3 by Salvatore Ruggiero and his colleagues describes the uptake of CE initiatives in four Baltic Sea region countries. (Estonia, Latvia, Lithuania and Poland). Chapter 4 by Nikolaos Ziozas and Theocharis Tsoutsos discusses REC initiatives in Greece using a Strategic Niche Management framework perspective. Chapter 5 by Chiara Candelise and Gianluca Ruggieri addresses the question how RECs in Italy evolved, and discusses recent experiences with the implementation of the Clean Energy Package in Italy. Chapter 6 by Lars Holstenkamp traces the development of CE in one of the front-running countries—Germany—from the early beginning as rural electrification cooperatives at the end of the nineteenth century to more recent developments. Chapters 7, 8, 9 and 10 address a number of more thematic issues around REC. In Chap. 7 Beau Warbroek addresses the importance of intermediaries and institutional support structures to empower RECs. This is based on an empirical study in the Fryslân region in the Netherlands. In Chap. 8 Frans Coenen and Thomas Hoppe address ECs seeking to get their members involved in energy-saving behaviors. The study presents survey results from five continental European countries (Spain, Italy, Denmark, Belgium and France). In Chap. 9 Mariëlle Feenstra and Florian Hanke take an inclusive CE perspective, focussing on gender in the CE movement. In Chap. 10, Lidewij Tummers discusses co-housing projects as a special form of CE, where theoretical approaches and empirical practices are useful to RECs and may expand their portfolios. Chapters 9 and 10 address the emerging attention to inclusiveness as a value in the community energy movement, and the energy transitions challenge at large. In the final chapter, the editors conclude the volume by answering the research questions, reflecting on the academic contributions, and more specifically discussing the four general observations made in the present chapter. Finally, suggestions for future research are presented.

References

Arentsen, M., & Bellekom, S. (2014). Power to the People: Local Energy Initiatives as Seedbeds of Innovation? *Energy, Sustainability and Society, 4*(1), 2. https://doi.org/10.1186/2192-0567-4-2

Bauwens, T. (2016). Explaining the Diversity of Motivations Behind Community Renewable Energy. *Energy Policy, 93*(C), 278–290.

Bauwens, T., & Defourny, J. (2017). Social Capital and Mutual Versus Public Benefits: The Case of Renewable Energy Cooperatives. *Annals of Public and Cooperative Economics, 88*(2), 203–232. https://doi.org/10.1111/apce.12166

Bauwens, T., Gotchev, B., & Holstenkamp, L. (2016). What Drives the Development of Community Energy in Europe? The Case of Wind Power Cooperatives. *Energy Research & Social Science, 13*, 136–147. https://doi.org/10.1016/j.erss.2015.12.016

Becker, S., Kunze, C., & Vancea, M. (2017). Community Energy and Social Entrepreneurship: Addressing Purpose, Organisation and Embeddedness of Renewable Energy Projects. *Journal of Cleaner Production, 147*, 25–36. https://doi.org/10.1016/j.jclepro.2017.01.048

Bomberg, E., & McEwen, N. (2012). Mobilizing Community Energy. *Energy Policy, 51*, 435–444. https://doi.org/10.1016/j.enpol.2012.08.045

Boon, F. P., & Dieperink, C. (2014). Local Civil Society Based Renewable Energy Organisations in the Netherlands: Exploring the Factors that Stimulate Their Emergence and Development. *Energy Policy, 69*, 297–307. https://doi.org/10.1016/j.enpol.2014.01.046

Brummer, V. (2018). Community Energy – Benefits and Barriers: A Comparative Literature Review of Community Energy in the UK, Germany and the USA, the Benefits It Provides for Society and the Barriers It Faces. *Renewable and Sustainable Energy Reviews, 94*, 187–196. https://doi.org/10.1016/j.rser.2018.06.013

Candelise, C., & Ruggieri, G. (2020). Status and Evolution of the Community Energy Sector in Italy. *Energies, 13*(8), 1888.

Cobutt, L. (2021). *Renewable Energy Communities in the European Energy Policy: An Evidence of Clean Energy for All Europeans, Not by All* (Egmont Paper No. 111).

Coenen, F. H. J. M., Hoppe, T., Chalkiadakis G., Tsoutsos, T., & Akasiadis C. (2017). *Exploring Energy Saving Policy Measures by Renewable Energy Supplying Cooperatives (REScoops)*. In Proceedings of the ECEEE.

Community Energy England. (2020). *Community Energy State of the Sector*. https://communityenergyengland.org/files/document/385/1592215769_CommunityEnergy-StateoftheSector2020Report.pdf

Community Energy Scotland. (2021). https://www.communityenergyscotland.org.uk/. Accessed 19 Apr 2021.

Creamer, E., Taylor Aiken, G., van Veelen, B., Walker, G., & Devine-Wright, P. (2019). Community Renewable Energy: What Does It Do? Walker and Devine-Wright (2008) Ten Years On. *Energy Research & Social Science, 57*, 101223. https://doi.org/10.1016/j.erss.2019.101223

Dóci, G., & Vasileiadou, E. (2015). "Let's do it ourselves" Individual Motivations for Investing in Renewables at Community Level. *Renewable and Sustainable Energy Reviews, 49*, 41–50. https://doi.org/10.1016/j.rser.2015.04.051

Dóci, G., Vasileiadou, E., & Petersen, A. C. (2015). Exploring the Transition Potential of Renewable Energy Communities. *Futures, 66*, 85–95. https://doi.org/10.1016/j.futures.2015.01.002

Gorroño-Albizu, L., Sperling, K., & Djørup, S. (2019). The Past, Present and Uncertain Future of Community Energy in Denmark: Critically Reviewing and Conceptualising Citizen Ownership. *Energy Research & Social Science, 57*, 101231. https://doi.org/10.1016/j.erss.2019.101231

Hewitt, R. J., Bradley, N., Baggio Compagnucci, A., Barlagne, C., Ceglarz, A., Cremades, R., & Slee, B. (2019). Social Innovation in Community Energy in Europe: A Review of the Evidence. *Frontiers in Energy Research, 7*(31). https://doi.org/10.3389/fenrg.2019.00031

Hitchins, T., Pauly, R., De Rynck, F., & Verschuere, B. (2020). Energiecooperaties en lokale besturen in Vlaanderen. Casestudie rapport, Januari 2020, Universiteit Gent, Flemish Strategic Basic Research (SBO).

Hoffman, S. M., & High-Pippert, A. (2010). From Private Lives to Collective Action: Recruitment and Participation Incentives for a Community Energy Program. *Energy Policy, Elsevier, 38*(12), 7567–7574.

Hoppe, T., & Coenen, F. H. J. M. (2016). *Exploring Interventions and Tools Used by REScoops to Lower Householders' Energy Consumption and Stimulate Investment in RES Projects*. Annual NIG Work Conference 2016.

Hoppe, T., Coenen, F. H. J. M., & Bekendam, M. T. (2019). Renewable Energy Cooperatives as a Stimulating Factor in Household Energy Savings. *Energies, 12*(7), 1188.

Hoppe, T., Graf, A., Warbroek, B., Lammers, I., & Lepping, I. (2015). Local Governments Supporting Local Energy Initiatives: Lessons from the Best

Practices of Saerbeck (Germany) and Lochem (The Netherlands). *Sustainability, 7*(2), 1900–1931.
Hoppe, T., van der Vegt, A., & Stegmaier, P. (2016). Presenting a Framework to Analyze Local Climate Policy and Action in Small and Medium-Sized Cities. *Sustainability* (Switzerland), *8*(9), 847. https://doi.org/10.3390/su8090847.
Horstink, L., Wittmayer, J. M., & Ng, K. (2021). Pluralising the European Energy Landscape: Collective Renewable Energy Prosumers and the EU's Clean Energy Vision. *Energy Policy, 153*, 112262. (The Netherlands) *Sustainability, 7*(2), 1900–1931.
Hufen, J. A. M., & Koppenjan, J. F. M. (2015). Local Renewable Energy Cooperatives: Revolution in Disguise? *Energy. Sustainability and Society, 5*(1), 18. https://doi.org/10.1186/s13705-015-0046-8
International Co-operative Alliance (ICA). (2017). https://www.ica.coop/en/cooperatives/cooperative-identity. Accessed 19 Apr 2021.
Kooij, H.-J., Oteman, M., Veenman, S., Sperling, K., Magnusson, D., Palm, J., & Hvelplund, F. (2018). Between Grassroots and Treetops: Community Power and Institutional Dependence in the Renewable Energy Sector in Denmark, Sweden and the Netherlands. *Energy Research & Social Science, 37*, 52–64. https://doi.org/10.1016/j.erss.2017.09.019
Lafferty, W. M., & Coenen, F. H. J. M. (2001). Conclusions and Perspectives. In W. M. Lafferty (Ed.), *Sustainable Communities in Europe* (pp. 266–304). Earthscan.
Oteman, M., Kooij, H.-J., & Wiering, M. (2017). Pioneering Renewable Energy in an Economic Energy Policy System: The History and Development of Dutch Grassroots Initiatives. *Sustainability, 9*(4), 550. https://doi.org/10.3390/su9040550
Oteman, M., Wiering, M., & Helderman, J.-K. (2014). The Institutional Space of Community Initiatives for Renewable Energy: A Comparative Case Study of the Netherlands, Germany and Denmark. *Energy, Sustainability and Society, 4*(1), 11. https://doi.org/10.1186/2192-0567-4-11
REScoop.eu. (2018). *A Landmark Day for Europe's March Towards Energy Democracy; European Parliament and Council Reach Agreement on the Renewables Directive*. Retrieved from https://www.rescoop.eu/blog/a-landmark-day-for-europe-s-march-towards-energy-democracy
Rescoop.eu/network. (2021). Available online: https://www.rescoop.eu/network. Accessed 19 Apr 2021.
Roberts, J. Charter (2020). Power to the People? Implications of the Clean Energy Package for the Role of Community Ownership in Europe's Energy

Transition. *Review of European, Comparative & International Environmental Law, 29*(2), 232–244. https://doi.org/10.1111/reel.12346
Schwencke, A. M. (2017). *Lokale energiemonitor 2017*. Retrieved from Utrecht: https://www.hieropgewekt.nl/uploads/inline/Lokale_Energie_Monitor_2017_3.pdf
Schwencke, A. M. (2021). *Lokale Energie Monitor 2020*. Retrieved from The Hague: https://www.hieropgewekt.nl/lokale-energie-monitor
Seyfang, G., & Haxeltine, A. (2012). Growing Grassroots Innovations: Exploring the Role of Community based Initiatives in Governing Sustainable Energy Transitions. *Environment and Planning C: Government and Policy, 30*(3), 381–400. https://doi.org/10.1068/c10222
Seyfang, G., Hielscher, S., Hargreaves, T., Martiskainen, M., & Smith, A. (2014). A Grassroots Sustainable Energy Niche? Reflections on Community Energy in the UK. *Environmental Innovation and Societal Transitions, 13*, 21–44. https://doi.org/10.1016/j.eist.2014.04.004
Soeiro, S., & Dias, M.F. (2019). *Renewable Energy Cooperatives: A Systematic Review*. Paper Presented at the 2019 16th International Conference on the European Energy Market (EEM), 18–20 September 2019.
Union renovables coop. (2021). http://www.unionrenovables.coop/grupo/. Accessed 19 Apr 2021.
Van der Horst, D. (2008). Social Enterprise and Renewable Energy: Emerging Initiatives and Communities of Practice. *Social Enterprise Journal, 4*(3), 171–185. https://doi.org/10.1108/17508610810922686
Van der Schoor, T., & Scholtens, L. J. R. (2015). Power to the People: Local Community Initiatives and the Transition to Sustainable Energy. *Renewable and Sustainable Energy Reviews, 43*, 666–675. https://doi.org/10.1016/j.rser.2014.10.089
Van der Schoor, T., & Scholtens, L. J. R. B. (2021). *Scientific Approaches of Community Energy, a Literature Review*. Centre for Energy Economics Research (CEER).
Van Veelen, B., & van der Horst, D. (2018). What Is Energy Democracy? Connecting Social Science Energy Research and Political Theory. *Science Energy Research & Social, 46*, 19–28.
Warbroek, B., & Hoppe, T. (2017). Modes of Governing and Policy of Local and Regional Governments Supporting Local Low-Carbon Energy Initiatives; Exploring the Cases of the Dutch Regions of Overijssel and Fryslân. *Sustainability* (Switzerland), *9*(1), 75. https://doi.org/10.3390/su9010075.

Warbroek, B., Hoppe, T., Coenen, F., & Bressers, H. (2018). The Role of Intermediaries in Supporting Local Low-Carbon Energy Initiatives. *Sustainability, 10*(7), 2450.

Wierling, A., Schwanitz, V. J., Zeiß, J. P., Bout, C., Candelise, C., Gilcrease, W., & Gregg, J. S. (2018). Statistical Evidence on the Role of Energy Cooperatives for the Energy Transition in European Countries. *Sustainability, 10*(9), 3339.

Yadoo, A., & Cruickshank, H. (2010). The Value of Cooperatives in Rural Electrification. *Energy Policy, 38*(6), 2941–2947. https://doi.org/10.1016/j.enpol.2010.01.031

Yildiz, Ö., Rommel, J., Debor, S., Holstenkamp, L., Mey, F., Müller, J. R., & Rognli, J. (2015). Renewable Energy Cooperatives as Gatekeepers or Facilitators? Recent Developments in Germany and a Multidisciplinary Research Agenda. *Energy Research & Social Science, 6*, 59–73. https://doi.org/10.1016/j.erss.2014.12.001

2

"What Are Energy Communities Under the EU's Clean Energy Package?"

Joshua Roberts

2.1 Introduction

Community ownership of energy resources has come a long way since the early twentieth century when cooperatives constructed distributed electricity and heating networks, some of which persist to this day. Followed by citizen movements in the 1970s and 1980s in Denmark, Germany and the Netherlands that aimed to develop renewable energy technologies, community ownership is now acknowledged in EU policy as a precondition for a successful and equitable clean energy transition.

The author would like to acknowledge that the material in this chapter that explains the energy community definitions under the Clean Energy Package were drawn from "Energy Communities Under the Clean Energy Package: Transposition Guidance", a report prepared by REScoop.eu and ClientEarth intended to provide Member States with guidance on how to draft new EU rules on energy communities into national legal frameworks.

J. Roberts (✉)
REScoop.eu, Brussels, Belgium
e-mail: josh.roberts@rescoop.eu

© The Author(s), under exclusive license to Springer Nature Switzerland AG 2021
F. H. J. M. Coenen, T. Hoppe (eds.), *Renewable Energy Communities and the Low Carbon Energy Transition in Europe*, https://doi.org/10.1007/978-3-030-84440-0_2

In May 2019, the European Union (EU) institutions concluded the final legislative files for the Clean Energy for All Europeans Legislative Package (CEP), a legal framework that will help the EU meet its 2030 climate and energy objectives. In particular, the recast Directive 2018/2001 (Renewable Energy Directive II, or REDII),[1] recast Directive 2019/944 (the Internal Electricity Market Directive, or IEMD)[2] define two new concepts labelled 'renewable energy communities' (RECs) and 'citizen energy communities' (CECs). Furthermore, together with the recast Regulation 2019/943 (the Internal Electricity Market Regulation, or IEMR),[3] the REDII and IEMD contain provisions that establish a supportive EU legal framework for community ownership. Specifically, they require Member States to secure certain rights of energy communities and establish enabling frameworks to ensure a level playing field and promote their development.

These new REC and CEC definitions attempt to consolidate the characteristics of already existing community energy initiatives, while trying to maintain flexibility so that new organisational models can develop according to national legal contexts. However, decision-makers will face a number of challenges as they write these new definitions into national law. Furthermore, these challenges will not dissipate once these definitions are implemented, and their use by citizens and other energy market actors will ultimately determine utility in driving social innovation and in democratising the energy sector. All of these developments will need to be carefully monitored.

This chapter will describe the characteristics of energy communities as they are defined under the CEP. First, it will briefly explain the European Commission's (Commission) rationale for including energy communities in the CEP. Second, it will explain each of the principles that are included

[1] Directive (EU) 2018/2001 on the promotion of the use of energy from renewable sources (recast) [2018] OJ
L328/82 (RED II).

[2] Directive (EU) 2019/944 on common rules for the internal market for electricity and amending Directive
2012/27/EU (recast) [2019] OJ L158/125 (IEMD).

[3] Regulation (EU) 2019/943 on the internal market for electricity (recast) [2019] OJ L158/54 (IEMR).

in the REC and CEC definitions, comparing and contrasting them where relevant. Third, the chapter will identify issues that national decision-makers currently face as they draft these new definitions into their national legal frameworks. Finally, this chapter concludes with observations on the prospects of these new legal concepts to further support community ownership and energy democracy in Europe's energy transition.

2.2 Why Include Energy Communities in the CEP?

Historically, community energy initiatives have been used by local citizens as a tool of social innovation to achieve many different goals according to local priorities. Correspondingly, legal concepts of community energy have mainly developed according to local and national socio-political and legal contexts.

In 2015, for several reasons community energy started gaining attention from EU policy. First, in its Energy Union Strategy, the Commission acknowledged the value of community energy towards helping meet national and EU renewables objectives and committed to putting citizens 'at the centre' of the Energy Union.[4] Second, it became apparent to the Commission that the integration of renewables into the European internal energy market (IEM) has negatively impacted community energy projects. For instance, Directorate-General (DG) Competition's 2014 revision of its Guidelines on State Aid for Environmental Protection and Energy (EEAG),[5] and subsequent changes to national renewables support schemes, have pushed community projects out of the market.[6]

In the recitals to the REDII and IEMD, the Commission makes its case for supporting energy communities with EU policy. According to

[4] Commission (EU) 'A Framework Strategy for a Resilient Energy Union with a Forward-Looking Climate Change Policy' (Communication) COM(2015) 80 final, 25 February 2015, 2.

[5] Commission (EU) 'Guidelines on State Aid for Environmental Protection and Energy 2014–2020' [2014] OJ C200/1 (2014 EEAG).

[6] Commission (EU) 'Impact Assessment Accompanying the document Proposal for a Directive of the European Parliament and of the Council on the promotion of the use of energy from renewable sources' (Staff Working Document) SWD(2016) 418 final, 30 November 2016, 32.

the recitals of the REDII, community energy provides added value in terms of local acceptance of renewable energy, access to additional private capital and further local investment, more choice for consumers and greater participation by citizens in the energy transition.[7] The IEMD adds that community energy helps promote inclusion, uptake of new technologies and consumption patterns that can deliver flexibility to the energy system, provision of affordable renewable energy to its members, and advancement of energy efficiency and the fight against energy poverty.[8]

Furthermore, the recitals to the REDII and IEMD acknowledge the unequal position of energy communities vis-à-vis other actors in the market. Specifically, the REDII states that "the specific characteristics of local renewable energy communities in terms of size, ownership structure and the number of projects can hamper their competition on an equal footing with large-scale players, namely competitors with larger projects or portfolios."[9] The IEMD echoes this, stating that "citizen energy communities constitute a new type of entity due to their membership structure, governance requirements and purpose."[10]

These high-level acknowledgments are quite important, as they provide the policy basis for the energy community definitions in the CEP, as well as requirements for Member States to establish enabling frameworks and ensure specific rights of energy communities that flow from their acknowledgment as actors in Europe's IEM.

2.3 What Are Energy Communities Under the Clean Energy Package?

From the recitals of the REDII and IEMD, energy communities can be identified based on certain characteristics such as size, purpose, ownership and membership, and internal governance. As such, under the CEP energy communities are meant to be understood as an organisational

[7] REDII, Recital 70.

[8] IEMD, Recital 43.

[9] REDII, Recital 71.

[10] IEMD, Recital 46.

concept.[11] The main aim of energy communities is to self-organise around an energy-related activity (e.g., generation and sale of renewable energy) in order to provide services or other socio-economic and environmental benefits to the members and/or the local community.

The REDII and the IEMD contain two new definitions acknowledging the ability of citizens to organise through energy communities in order to take ownership and responsibility for different activities across the energy market:

- 'Renewable Energy Community' (REC)—defined in the REDII; and
- 'Citizen Energy Community' (CEC)—defined in the IEMD (Table 2.1).

Table 2.1 Energy Community definitions in the CEP

Article 2(16) Recast Renewable Energy Directive 'Renewable Energy Community'	Article 2(11) Recast Electricity Directive 'Citizen Energy Community'
A legal entity: Which, in accordance with the applicable national law, is based on open and voluntary participation, is autonomous, and is effectively controlled by shareholders or members that are located in the proximity of the renewable energy projects that are owned and developed by that legal entity; the shareholders or members of which are natural persons, SMEs or local authorities, including municipalities; the primary purpose of which is to provide environmental, economic or social community benefits for its shareholders or members or for the local areas where it operates, rather than financial profits.	A legal entity that: Is based on voluntary and open participation and is effectively controlled by members or shareholders that are natural persons, local authorities, including municipalities, or small enterprises; has for its primary purpose to provide environmental, economic or social community benefits to its members or shareholders or to the local areas where it operates rather than to generate financial profits; and may engage in generation, including from renewable sources, distribution, supply, consumption, aggregation, energy storage, energy efficiency services or charging services for electric vehicles or provide other energy services to its members or shareholders

[11] Commission (EU) Presentation on "Energy Communities: implementation of the Clean Energy Package" as informal advice to Member States on existing knowledge and examples, p. 13. On file with author.

The starting point for both definitions is the establishment of a legal entity. Furthermore, the legal entity must be organised around specific ownership and governance principles, and a non-commercial purpose. Interestingly enough, some of these principles are based on the seven International Co-operative Alliance (ICA) Principles of cooperatives, also known as the 'Rochdale Principles'.[12] Similar to the ICA Principles, the principles contained in the energy community definitions can be integrated into any legal form.

Together, the elements of both definitions convey a similar concept: a particular way to organise collective ownership around a particular energy-related activity. Therefore, many of the elements in the REC and CEC definitions are identical, or very similar.

Nevertheless, there are also material differences between the REC and CEC definitions. These differences have two justifications. First, the Commission has stated that the objectives of the provisions on RECs and CECs in the CEP are different. For RECs, the provisions must be seen as a way to help meet the European renewable energy target. Therefore, RECs have a more local focus (in line with promoting acceptance to renewables) and are intended to receive more benefits or incentives so that they can develop at national level.[13] They also have more stringent participation and governance requirements to limit the impact on competition and avoid abuse. On the other hand, CECs are acknowledged as a new market actor in the electricity sector. Therefore, the aim is to provide a level playing field.[14]

From a sectoral perspective, CECs and RECs also differ slightly. As the IEMD focuses on the electricity sector, CECs do not have a technology-specific focus. In contrast, given the REDII's focus on renewable energy sources, RECs are limited to activities related to non-fossil sources of energy. In reality, as most energy communities focus on renewable sources of electricity, there is significant overlap between these two definitions. Therefore, while it might be easy to simply describe RECs as a subset of

---

[12] Ibid., p. 20; *See also* International Co-operative Alliance (ICA). Cooperative identity, values and principles. Available at: https://www.ica.coop/en/cooperatives/cooperativeidentity

[13] Commission (EU), n. 12 at p. 10.

[14] Ibid.

CECs due to their narrower sectoral focus and more stringent participation requirements, this description is an over-generalisation. This has led to complications in the process of writing the definitions into national law, and will be dealt with later in the chapter.

The following sections will be dedicated to elaborating each of the principles that must be encompassed within a particular initiative that engages citizens to have it acknowledged as a REC or CEC, and what they mean.

2.4 Principles

2.4.1 Legal Form

Both RECs and CECs must be a legal entity. This means that a citizen initiative that does not have a legal form would not be eligible to be recognised as a REC or CEC. Member States have discretion to choose the form of legal entity that an energy community may use. This is confirmed in the recitals of the REDII and the IEMD.[15]

Today, several legal forms are used to set up energy communities: cooperatives, different forms of partnerships, companies with a community interest, foundations, non-profit customer-owned enterprises, various forms of social enterprises, and associations (e.g., housing) (Roberts et al., 2014).[16] Certain forms of municipal or public ownership may also be considered an energy community. Member States have discretion to adopt the legal forms that they feel are appropriate. This discretion is limited only by a requirement that any legal forms still need to meet the eligibility criteria contained in the EU level definitions. This possibility leaves open the potential for new legal forms to be utilised for energy communities in the future.

[15] REDII, recital 71; and IEMD, recital 44.

[16] For examples of different legal entities that are currently in use for local renewable community energy initiatives in the UK, Denmark and Germany, see J. Roberts, F. Bodman and R. Rybski (2014). Community Power: Model legal frameworks for citizen-owned renewable Energy (ClientEarth), p. 34.

2.4.2 The Primary Purpose of an Energy Community

RECs and CECs are not intended to be created for commercial purposes. In this regard, the text of the definitions in the REDII and the IEMD are identical. The REDII and the IEMD both state that the primary purpose of RECs and CECs are "to provide environmental, economic or social community benefits for its shareholders or members or for the local areas where it operates, rather than financial profits."[17] The definitions prohibit RECs or CECs from adopting a commercial motive (i.e., selling a product or service with the primary purpose of generating profits that are then distributed to their owners/shareholders). This is a significant characteristic that distinguishes energy communities from initiatives that might be set up by a traditional energy company as a commercial service to consumers.

The language of the directives does not forbid energy communities from generating profits per se. Profits, which energy communities need for social innovation, may be reinvested into the community's activities (e.g., renewables generation projects) or be used to pursue other public interest aims such as local development, education, or solidarity programmes (e.g., tackling energy poverty).

Furthermore, the definition does not prohibit RECs or CECs from providing a return on investment to the members. However, returns on investment and other financial benefits to members should be secondary to other general aims of the community. This would, for instance, exclude industrial or commercial consumers from setting up a REC or CEC to reduce energy costs, because such costs impact their bottom line.

The benefits generated by RECs and CECs should also be aimed at one of two groups: shareholders/members, or the local community. First, benefits may be provided to the shareholders or members of the energy community. This is very typical of cooperatives,[18] and in the energy context it usually applies to provision of energy supply or flexibility services,

[17] REDII, Article 2(16); and IEMD, Article 2(11).

[18] For instance, national legislation usually requires that activities are performed for the mutual benefit of members, meaning provision of services or return on investment. *See* Court of Justice of the EU (CJEU) Joined Cases C-78/08 to C-80/08 Paint Graphos Soc coop. Alr., ECLI:EU:C:2011:550.

improving awareness of climate change, training on rationalising energy use, or return on investment from a renewable energy project. Second, benefits can be aimed at the local community in which the REC or CEC operates. This reflects the aims of many local social enterprises, which often provide services and benefits to local citizens.

The types of benefits that RECs and CECs can provide their members and the community are quite broad, encompassing economic, environmental or social benefits. These could encompass, for example (but not limited to):

- a return on investment to members;
- promotion of local development;
- investment in public infrastructure;
- reduced energy bills—at least for households and other non-professional customers;
- increased production of locally developed renewable energy;
- greenhouse gas emissions reductions (e.g., carbon dioxide, or CO_2);
- provision of different services (e.g., energy supply, sharing, advice) to members;
- investment in energy efficiency, energy poverty/solidarity initiatives;
- education and training for members, school children, and/or the broader public;
- promotion of energy democracy; and
- citizen empowerment.

As such, the directives provide space for citizens to use energy communities for social innovation.

2.4.3 Open and Voluntary Participation

Under the REDII and IEMD, RECs and CECs must be "based on open and voluntary participation."[19] These are broad principles that require further specification at national level.

---

[19] REDII, Art 2(16); and IEMD, Art 2(11).

The recitals to the REDII provide some guidance on the meaning of openness. Recital 71 states that "participation in [RECs] should be open to all potential local members based on objective, transparent and non-discriminatory criteria." This means that as long as they accept new members, RECs and CECs should not be able to use arbitrary criteria or exclude eligible citizens that are willing to undertake the responsibilities of membership (e.g., exorbitant entry fees). This should not prohibit RECs from developing criteria for membership (e.g., minimum investment thresholds), as long as they are not arbitrary or meant to discriminate.

However, openness could also be interpreted more broadly as a way to ensure diversity in the energy community. For instance, energy communities may want to encourage households that do not have sufficient disposable income to be able to become members, or at least be eligible to receive benefits derived from the community's activities.[20] This can be achieved by reducing or even disposing away with requirements to contribute capital to the community in order to become a member, or even providing services to non-members, excusing membership entirely. This is supported by the fact that in their enabling frameworks, Member States must ensure that participation in RECs is accessible to all consumers, including those in low-income or vulnerable households.[21]

Under the REDII and IEMD, voluntariness is framed around ensuring the right of members or shareholders to leave the energy community. Here, two potential relationships between the energy community and its participants are relevant: the business–customer relationship, and the member/investor relationship.

Regarding the business–customer relationship, under EU law consumers must be able to maintain their rights as a consumer, which includes the right to switch suppliers. Therefore, no one can be forced to join or to stay in an energy community, and normal rules regarding consumers' rights to switch suppliers or service provider (e.g., aggregator) must be

[20] COMPILE (2021). The Energy Community Toolkit. Deliverable 4.1, COMPILE Project, p. 94. https://www.compile-project.eu/wp-content/uploads/COMPILE_D4_1_The_Energy_Community_Toolkit_FINAL.pdf

[21] REDII, Art 22(4)(f).

respected.[22] However, this requirement does not apply to the relationship between consumers and energy communities as a distribution system operator (e.g., electricity or heating and cooling), as there is no consumer right to choose your network operator. In such situations, the customer would still have the right to switch supplier or service provider, but not the network itself.

With regard to the investor relationship between a REC or CEC and its member or shareholders, national laws around companies, associations, or cooperatives already cover processes for joining or leaving an enterprise. As such, these rules will differ depending on which legal entities Member States choose to use as a basis for energy communities.

2.4.4 Eligibility to Be Member or Stakeholder in an Energy Community

Regarding eligibility to participate in an energy community, the REC and CEC definitions diverge. The REDII explicitly identifies which actors are eligible to participate in a REC:

- natural persons;
- small and medium enterprises (SMEs); or
- local authorities (including municipalities).[23]

The REDII limits which companies can be a member of a REC, based on size. Specifically, large companies, which are defined under the EU Commission's Recommendation Concerning the Definition of Micro, Small and Medium-sized Enterprises, are prohibited from becoming a member or shareholder in an REC.[24]

Article 22(1) of the REDII also allows Member States to prohibit or place limitations on the ability of energy companies to be a member in a REC. The rationale for providing Member States with such discretion is

[22] REDII, Art 22(4)(b); and IEMD, Art 16(1)(a).

[23] REDII, Art 2(16).

[24] Commission (EU) Recommendation concerning the definition of micro, small and medium-sized enterprises, OJ L 124, 20.05.2003, p 36 (SME Recommendation).

to prevent abuse of RECs by energy companies that might have a commercial motive for setting up energy communities or providing them with services. This would not prohibit a company from entering into a partnership agreement with a REC, for instance through a bilateral contractual arrangement, or from providing services to the REC.

In contrast, the IEMD does not place any limits on eligibility to become a member or shareholder of a CEC. This is supported by the recitals, which state that "membership of citizen energy communities should be open to all categories of entities."[25] This means that large enterprises—not just SMEs—may become members of a CEC.

2.4.5 Effective Control

In order to ensure energy communities are driven by local actors, and to prevent abuse by commercial market actors, the REDII and IEMD elaborate requirements for which members or shareholders can effectively control the community.

'Effective control' is not defined in the CEP. However, control refers generally to a situation in which a particular member or shareholder within a legal entity (company, natural person, or local authority) wields significant influence over the management or decision-making situation, based on their voting power or shares held. In other words, a company is 'controlled' if there is a group of shareholders that bring together enough shares (e.g., a majority, or a significant minority) to give that group a decisive voice in managing the company.[26]

The EU Commission has indicated that it is for Member States to define what 'effective control' means in this context, consistent with their existing legislation on corporate governance (i.e., company law), tax and competition.[27] This means that effective control will differ between each Member State, ranging from but not limited to:

[25] IEMD, Recital 44.

[26] CEOpedia. https://ceopedia.org/index.php/Controlled_company

[27] Commission (EU), n. 12 at p. 21.

- having majority voting rights in general meetings of the REC through the amount of capital held (e.g., 51 per cent) or by virtue of an agreement with other members or shareholders;
- having a decisive influence in the decisions of the REC by virtue of the shares held or its position on the board; or
- when it has enough voting power or shares to wield veto power against the other members or shareholders.

RECs and CECs are different in their rules for which members or shareholders are allowed to exercise effective control of the community. For CECs, effective control is limited based on the size of the member or shareholder, while for RECs effective control must be local.

RECs must be "effectively controlled by shareholders or members that are located in the proximity of the renewable energy projects that are owned and developed by that legal entity." Proximity is not defined anywhere and is therefore left to further interpretation at the national level. Nevertheless, given that this requirement exists to promote public acceptance of new renewable energy projects, it should be generally understood as the geographical scope in which the REC is located and has an impact. Member States can use their own methods and criteria to define proximity using existing land use planning or zoning documents, the use of municipal boundaries, or other criteria (e.g., consultation requirements).

The directives do not define proximity in relation to any specific activity that the REC might perform. However, it is important to understand that the number of people affected by a particular REC initiative will differ from one activity to another. For instance, a self-consumption or energy sharing initiative will likely take place on a rather limited geographical scale, while supply or provision of flexibility services may indicate that the energy community operates according to the size of the market (e.g., regional or national).

CECs are not restricted by locality. Instead, limits on effective control are established according to the size of the member or shareholder. Specifically, the IEMD restricts effective control to natural persons, local authorities and small and micro-enterprises.[28] The recitals justify this restriction based on the idea that the "decision-making powers within a CEC should be limited to members or shareholders that are not engaged in largescale commercial activity and for which the energy sector does not constitute a primary area of economic activity."[29] This reinforces the idea that energy communities are initiatives self-organised by non-professional actors in the energy market.

This recital has received varying interpretations regarding the limitations that should be placed on energy companies in the decision-making of the CEC. If relying on the recital, all large companies could be prohibited from taking any part in any decision-making in the CEC. However, according to the definition itself, which is the legally operative text, medium-sized and large companies (both inside and outside the energy sector) could take part in decision-making of the CEC, as long as effective control or direction of the decision-making of the CEC remains with the other members.

Given that there are no limits to eligibility for participation in CECs, there is potentially more risk that CECs are abused by larger commercial and industrial customers, as well as traditional commercial energy companies. Furthermore, because restrictions on effective control are based on size, it is also possible that companies will create special purpose legal vehicles that qualify as 'small' in order to get around this rule.

2.4.6 Autonomy to Ensure Democratic Governance

Autonomy is an additional governance requirement that only applies to RECs. Under the REDII, RECs must be autonomous. While autonomy is not defined by the REDII, the recitals provide guidance on its meaning: "to avoid abuse and to ensure broad participation, REC should be capable of remaining autonomous from individual members and other traditional market actors that participate in the community as members

[28] According to Article 2(8) of the REDII, a micro, small, and medium-sized enterprise (SME) is defined in Annex 2 of Commission (EU) Recommendation concerning the definition of micro, small and medium-sized enterprises, OJ L 124, 20.05.2003, p 36 (SME Recommendation).

[29] IEMD, recital 44.

or shareholders, or who cooperate through other means such as investment."[30]

Under this context, autonomy can be seen as an extension of the principle of effective control, or rather, a limit on effective control being exercised by individual members. Therefore, the requirement that a REC be autonomous is meant to ensure that the community remains controlled jointly by the members as a collective, rather than by a single member or a small group of members. Furthermore, because it is an extension of effective control, Member States have discretion to also apply the autonomy requirement to CECs.

The principle of autonomy within the REDII derives strongly from the ICA Principles. Under the ICA, autonomy supports democratic internal decision-making so that all members are adequately represented (regardless of their amount of investment).[31] Autonomy is also about guaranteeing economic and financial autonomy, meaning that business relationships with traditional market actors should not undermine the community's decision-making independence.[32] According to this principle, single members, or a category of members (e.g., investor or for-profit non-user members) should not control the cooperative, and all members as a whole should be free to govern the cooperative without external influences, for example, through third party contracts or financing arrangements (Fici, 2012).[33]

As Member States have significant discretion on how to apply this principle, autonomy can be reflected in national energy communities' legislation in a number of ways. For instance:

- Limitations may be placed on the amount of capital held by one member, or a group/class of members (e.g., for-profit enterprises), so that the collective autonomy of the REC is not endangered.

[30] REDII, recital 71.

[31] International Cooperative Alliance (ICA). "Guidance Notes to the Co-operative Principles," p. 2.

[32] Ibid., at p. 47.

[33] A Fici (2012). "Cooperative Identity and the Law," Eurisce Working Paper No. 23/12 (Euricse), p. 13.

- Limitations may be imposed on the withdrawal, or liquidation, of investment by individual members so that the REC's continued existence is not put at risk.
- Rules can limit sources of external financing (from non-user members or non-member users) that the REC can accept, which even without giving voting and/or participatory rights could limit the REC's autonomy.
- Rules may stipulate that decisions must be democratic and based on the one-person-one-vote principle. The one-person-one-vote principle can also have the added benefit of ensuring equal and non-discriminatory treatment between members or shareholders in the REC, which is a requirement under Article 22(4)(i) of the REDII.

Given that RECs and CECs envision participation by natural persons, municipalities and companies, it may be very difficult to ensure equal decision-making rights by all those involved. For instance, most individuals, depending on their investment experience or economic position, will not be in a place to take decisions on an equal playing field with more sophisticated members such as companies. To alleviate such situations, and to ensure open and democratic governance, it may be necessary to include veto mechanisms for minority members, or weighted voting for different categories of members, in their founding statutes. It may also be necessary to set up a representative that can speak for a specific class of shareholders, such as consumers. Such mechanisms might be especially relevant where there is a large number of vulnerable households participating in the community.

2.5 What Can Energy Communities Do?

Consistent with the emergence of energy community initiatives across different energy-related activities, the CEP ensures that RECs and CECs can participate across the energy sector. The IEMD is quite explicit in this regard. The CEC definition enumerates a comprehensive list of activities that a CEC can engage in, including:

- generation, including from renewable sources;
- distribution;
- supply;
- consumption;
- aggregation;
- energy storage;
- energy efficiency services;
- charging services for electric vehicles; or
- provide other energy services to its members or shareholders.

In contrast, the REC definition in the REDII does not list any activities. However, the activities that RECs are provided with rights to undertake are mentioned in Article 22. In particular, RECs are granted the right to produce, consume and sell renewable energy.[34] They are also guaranteed non-discriminatory treatment as final consumers, producers, suppliers, distribution system operators or as other market participants.[35] Therefore, it must be concluded that RECs are entitled to engage in any number of different activities, as long as those activities are linked with renewable energy.

Both directives also specifically mention that energy communities must be allowed to share renewable energy that is produced by the community.[36] Because this new activity has its own special mention in the articles on energy communities, many spectators have wrongly assumed that energy communities are specifically intended for this activity. The same assumption has been made about collective self-consumption, for which there are separate provisions in Article 21 of the REDII. This has resulted in many first-generation national definitions of energy communities that are framed around a specific activity such as collective renewables self-consumption of local energy sharing. This issue will be tackled more below.

When it comes to network ownership and operation, Member States have much more discretion whether to allow energy communities to

[34] REDII, Article 22(2)(a).

[35] REDII, Article 22(4)(e).

[36] REDII, Article 22(2)(b); and IEMD Article 16(3)(e).

perform this activity. The IEMD explicitly provides Member States with the option of whether or not to allow CECs to own, establish, purchase or lease electricity distribution networks and autonomously manage them.[37] The REDII does not contain any reference to the ability of RECs to manage district heating or cooling networks. Nevertheless, the REDII does require Member States to ensure that RECs are granted non-discriminatory treatment with regard to their activities, rights and obligations as distribution system operators.[38] Therefore, it is reasonable to conclude that if Member States allow, RECs should be able to operate district heating and cooling networks.

2.6 Emerging Challenges with the Transposition and Implementation of EU Energy Community Definitions at National Level

Given the novelty of energy communities as a legal concept, and the fact that the EU definitions have been composed as a set of broad principles, there are several challenges to writing these EU rules (also known as transposition) into national legal systems. This is not a huge surprise, given the already divergent views of what constitutes community energy in literature, and the fact that these concepts are now in the hands of national decision-makers who may only be hearing of energy communities for the first time.

Nevertheless, these first-generation energy community definitions at national level will have significant implications for the development of energy communities over the coming years. As such, it is important that decision-makers get it right; otherwise it could significantly delay the growth of energy communities, or they could be co-opted by commercial energy companies. Ultimately, this could undermine efforts to

[37] IEMD, Article 16(2)(b).

[38] REDII, Article 22(4)(e).

democratise the energy system and provide citizens with the ability to take ownership in Europe's energy transition.

Below are a number of issues that will impact the usefulness of energy communities at the local level.

2.6.1 Delayed and or Incomplete Transposition

A number of reports show that the pace of transposition of EU legislative provisions on energy communities is very uneven (Toporek & Provost, 2020; Standal & Aakre, 2021).[39] While some Member States are making good progress on the transposition of the CEP, others are only partially transposing the definitions. Germany has even gone as far to say that it does not currently plan to transpose energy community provisions of the CEP, including the definition, as it considers—wrongly—that it already has a sufficient legal framework for energy communities in place. Other Member States are delayed in transposition due to other circumstances such as upcoming national elections, and the redirection of resources towards dealing with COVID-19. The deadline for transposing the IEMD was 31 December, 2020, while the deadline for transposing the REDII was 30 June 2020. Ironically, more progress has been made to transpose provisions on RECs than for CECs. Regardless, it is clear that delays in the drafting of definitions and national legal frameworks for energy communities will create delays in their development.

2.6.2 Lack of Awareness of the Role Energy Communities Can Play in the Energy Transition

There is still a misconception, or a lack of understanding, about what activities and roles energy communities can perform across the energy sector. In particular, there has been a tendency by decision-makers and

[39] *See* Toporek, M. and Provost, L. (2020). Guidance for National Transposition of New EU Directives relating to Renewable Energy Prosumers. PROSEU—Prosumers for the Energy Union: Mainstreaming active participation of citizens in the energy transition (Deliverable No. 3.5); and Standal, K. and Aakre, S. (2021). Assessment Report on Technical, Legal, Institutional and Policy Conditions. COME RES—Advancing Renewable Energy Communities (Deliverable 2.1).

the traditional energy sector to view energy communities through the lens of a particular activity, such as collective self-consumption, energy sharing, or micro-grid projects. Lack of awareness by Member States was clearly visible when Member States submitted their draft National Energy and Climate Plans (NECPs), which they were required to complete to ensure that the EU meets its 2030 climate and energy targets (Roberts & Gauthier, 2019).[40] This lack of understanding has resulted in a number of Member States drafting (and even adopting) overly narrow definitions which only partially transpose the EU level definitions. While some Member States have started to correct early mistakes, these missteps have contributed to delays and a lack of legal clarity by those looking to use these new concepts.

2.6.3 Principles Getting Lost in Translation

Member States have broad discretion to interpret each of the participation and governance principles that are contained in the REC and CEC definitions. However, this creates a number of practical and legal challenges. First, experience so far has shown that many Member States are unaware of the ICA Principles, and are unsure how to further specify principles such as autonomy and openness in national legislation. This has led many Member States to simply copy-and-paste the national definitions into national law. Alternatively, ministries have delayed the process of further specifying these principles, requiring the development of by-laws or regulations. Again, this has resulted in delays towards providing necessary legal certainty for citizens, local authorities or companies interested in forming energy communities.

There is also a risk that in being translated into national law these definitions are appropriated and abused by well-resourced energy companies, and larger industrial and commercial energy consumers. Despite recitals text that clearly states these concepts were meant to benefit citizens, traditional market actors are already heavily lobbying national ministries to

[40] *See* Roberts, J. and Gauthier, C. (2019). "Energy Communities in the Draft National Energy and Climate Plans: Encouraging but Room for Improvements" (REScoop.eu; European University Viadrina).

tailor definitions for them. Given that the energy community labels are attached to special rights and privileges, if companies are successful in their lobby efforts they could create new market distortions, which would work against true citizen-led energy communities. If energy communities are seen as a 'citizen-washing' exercise, this could undermine support for the concept as a whole. National authorities will indeed be required to monitor and oversee the use of energy community definitions at national level.[41] However, if criteria on governance and participation are not further specified, it will be very difficult to monitor potential abuses.

2.6.4 Ensuring Coherence between RECs and CECs

When viewed together, it is fairly clear that both RECs and CECs are intended to reflect a common social or organisational concept allowing for collective ownership around different energy-related activities by different local actors. It would have been ideal if both EU definitions had contained identical or non-contradicting criteria. Unfortunately, both definitions contain minor differences that could make it difficult to ensure a coherent and clear relationship at national level (REScoop.eu and ClientEarth, 2020).[42]

While there are a few options for ensuring coherency between RECs and CECs, there is little likelihood that they will be pursued across all twenty-seven EU Member States. As such, there is potential for these concepts to diverge, even if from a technical perspective the overlap between RECs and CECs increases, which is highly likely as Europe moves toward sector coupling. For example, many RECs, in particular cooperatives, already operate district heating networks. As flexibility markets develop, RECs with such networks will be incentivised to provide flexibility to the power system, which is regulated under the IEMD. To ensure coherence and provide legal clarity, the differences between the REC and CEC definitions will likely need to be reconciled at EU level.

[41] IEMD, Article 59(z).

[42] *See* REScoop.eu and ClientEarth (2020). "Energy Communities Under the Clean Energy Package", p. 37.

2.6.5 Accommodation of Existing and New Legal Forms of Organising Energy Communities

While many energy communities exist using different legal forms, it is inevitable that some will fall outside the criteria that have been set by the EU level definitions. Furthermore, many decision-makers will need to assess which existing legal forms from their national company law are most appropriate to be used by RECs and CECs. Where energy communities already exist, Member States will still need to identify them and choose a set of legal forms that fit these existing citizen energy initiatives.

However, it is unclear whether national laws will be tailored to fit all existing legal forms. In particular, while cooperatives have been perhaps the most common and successful legal form utilised for energy communities, in some Member States there is animosity towards cooperatives. In other Member States that have opted for a more general approach, there is a lack of clarity regarding which existing legal forms might qualify as an energy community. For instance, in draft legislation on energy communities in Croatia, the government has proposed to allow legal entities that qualify as a non-profit organisation, creating questions about whether it will be possible to use cooperatives to create energy communities. Lastly, national decision-makers will need to identify additional legal forms that might be suitable for energy communities, such as home-owner associations and various forms of charities. In doing so, they will need to ensure that already existing legal entities can qualify as an energy community without too much administrative burden.

2.6.6 Balancing Inclusion Against the Potential for Competitive Abuse

As the REC and CEC definitions focus on specific legal entities, they do not address all of the different configurations of community ownership and collaboration with the traditional energy sector. In particular, there are many questions regarding whether and/or how to support softer forms of citizen energy initiatives, such as crowdfunding and co-ownership between communities and commercial companies. As long as Member States transpose the definitions in the REDII and the IEMD, they are free to come up with other energy community models.[43] However, there is a risk that adopting broader approaches could significantly water down the energy community concepts that have been defined under the CEP, or create the ability for commercial energy companies to undermine real citizen-led energy communities.

Abuse of energy communities in different Member States is already occurring. In Germany, for instance, the 2017 EEG (German support mechanism) established a definition and special rules for 'citizens' energy companies' for onshore wind tenders. In the first three rounds of bids, 97 per cent of successful bids came from eligible citizens energy companies. However, after assessing the individual projects in detail, it was shown that nearly all of these projects were established by bigger market players, and only eight projects could be considered a real citizen energy project (Pappa & Vanstintjan, 2020; RESCoop.eu, 2021).[44] Furthermore, due to abuses by large companies the Greek government recently withdrew many of the useful exemptions in licensing requirements for energy communities that it had included in its 2016 legislation.[45]

2.6.7 Accompanying Definitions with Adequate Rights and an Enabling Framework

There has been a great deal of attention on how to draft definitions of energy communities at national level, and in granting them the ability to perform different roles in the energy sector. However, less attention has been focused on ensuring that enabling frameworks are in place, particularly for RECs, and that that regulatory barriers are reduced so that energy communities can be integrated into the market. This has been a noticeable trend across almost all Member States that have undertaken the

[43] Commission (EU), n. 12 at p. 28.

[44] *See* Pappa, S. and Vanstintjan, D. (2020). "REScoop.eu Response to the Consultation on the EEAG and GBER revision."

[45] *See* REScoop.eu (2021). "Development of energy communities in Greece: Challenges and recommendations." Full report available in Greek.

transposition process so far. Currently, only one Member State, Ireland, has created a dedicated space for RECs so they can access its national renewables support scheme. Furthermore, while Portugal has written in its national legislation that the Ministry must undertake an assessment of the potential and barriers to the development of RECs at national level, no Member State has actually undertaken this assessment, even though it is a clear requirement in Article 22(4) of the REDII. Some Member States, such as Belgium and Austria, have introduced special regulatory exemptions for energy communities. Yet most decision-makers and regulators are still reluctant to grant energy communities many favours.

This is not to say that incremental progress is not being made. DG Competition has itself acknowledged that its existing EEAG need to be adjusted, fine-tuned or potentially aligned with new CEP rules, including treatment of self-consumption and energy communities in renewables support schemes.[46] Furthermore, the Council for European Energy Regulators (CEER) recently acknowledged that rules and administrative procedures for energy communities should be simple, "in order to ensure that these actors do not face discriminatory burdens or costs in comparison with established energy companies" (CEER, 2021).[47]

Nevertheless, much more progress in assessing the challenges, potential and drivers for RECs is needed across the EU. Otherwise, even the best energy community definitions are likely to go under-utilised. Not only might this result in a significant amount of legal challenges at national level, but it could place the spotlight on the EU Commission in terms of how much political (and legal) capital it is willing to use to ensure these provisions are properly transposed and faithfully implemented.

[46] Commission (EU). Commission staff working document, Fitness check of the 2012 State aid modernisation package, railways guidelines and short-term export credit insurance, Annex SWD (2020) 257 final, 30.10.2020, Annex 8, p 104.

[47] Council of European Energy Regulators (CEER) (2021). "Draft CEER Strategy 'Empowering Consumers for the Energy Transition'—Public Consultation Document (19 March 2021), p. 18.

2.7 Conclusions

As energy communities are a brand-new concept under EU law and in most Member States, writing these new EU rules into national legislation will be challenging. Many questions are currently being asked about what energy communities are, how to define them, what activities they should be able to participate in, how they should be regulated, and how their development should be supported.

When the CEP was finalised it was heralded as a game-changer for citizens wanting to get involved in the energy transition. As the transposition period expires, whereby Member States are required to translate these EU requirements into national law, the existing situation paints a more complicated picture. There is still a lack of awareness of the full potential of energy communities by national decision-makers, as well as a notable lack of ambition towards furthering their development. This is likely to result in an uneven playing field for energy communities across different Member States. It is also likely to put pressure on the Commission to get creative in using soft (e.g., support, finance) and potentially hard (e.g., enforcement) policy and legal tools to ensure Member States comply with their requirements.

Regardless of the pace of change, it is quite likely that energy communities themselves will also evolve as policy, law and regulation catches up to their development. Before EU legislation on energy communities existed, energy communities evolved from the bottom up according to local needs and national legal contexts. Now that EU policy exists, there will be a top-down development that will take place. This will inevitably change the existing landscape of what energy communities look like today. This trend is likely to be furthered by technological advancements and professionalisation inside existing energy communities. All of these drivers have the potential to impact the social innovation potential of energy communities, both positively and negatively.

One thing is clear. EU policy has begun a process of bringing energy communities into the mainstream. Where energy communities develop from here is as much dependent on policy makers as the actors that utilise them.

References

Council of European Energy Regulators (CEER). (2021, March 19). *Draft CEER Strategy. "Empowering Consumers for the Energy Transition"* (p. 18). Public Consultation Document. Ref: C21-SSG-05-03. Brussels.

Fici, A. (2012). *Cooperative Identity and the Law* (p. 13) (Eurisce Working Paper No. 23/12). Euricse.

Pappa, S., & Vanstintjan, D. (2020). REScoop.eu. Response to the Consultation on the EEAG and GBER Revision.

REScoop.eu (2021). *Development of Energy Communities in Greece: Challenges and Recommendations*. Full Report Available in Greek.

REScoop.eu and ClientEarth. (2020). *Energy Communities Under the Clean Energy Package* (p. 37).

Roberts, J., Bodman, F., & Rybski, R. (2014). *Community Power: Model Legal Frameworks for Citizen-Owned Renewable Energy* (p. 34). ClientEarth.

Roberts, J., & Gauthier, C. (2019). *Energy Communities in the Draft National Energy and Climate Plans: Encouraging but Room for Improvements*. REScoop.eu/European University Viadrina.

Standal, K., & Aakre S. (2021). *Assessment Report on Technical, Legal, Institutional and Policy Conditions* (Deliverable 2.1). COME RES – Advancing Renewable Energy Communities.

Toporek, M., & Provost, L. (2020). *Guidance for National Transposition of New EU Directives Relating to Renewable Energy Prosumers* (Deliverable No 3.5). PROSEU – Prosumers for the Energy Union: Mainstreaming Active Participation of Citizens in the energy transition.

3

Community Energy in the Eastern Baltic Sea Region: From Standstill to First Steps

Salvatore Ruggiero, Henner Busch, Aljosa Isakovic, and Teis Hansen

3.1 Introduction

In recent years, community energy (CE) has gained increasing attention for the role it can play in the decentralization of current energy systems and in the acceleration of transition towards a low-carbon society. Although the term CE is growing in popularity, there is not yet agreement about what it means. The recent EU renewable energy and electricity directives have introduced two new legal entities, called 'renewable energy communities' and 'citizen energy communities', that

S. Ruggiero (✉)
Center for Consumer Society Research, University of Helsinki, Helsinki, Finland

School of Business, Aalto University, Espoo, Finland
e-mail: salvatore.ruggiero@helsinki.fi

H. Busch
Department of Human Geography, Lund University, Lund, Sweden

Lund University Centre for Sustainability Studies, Lund, Sweden
e-mail: henner.busch@lucsus.lu.se

© The Author(s), under exclusive license to Springer Nature Switzerland AG 2021

F. H. J. M. Coenen, T. Hoppe (eds.), *Renewable Energy Communities and the Low Carbon Energy Transition in Europe*, https://doi.org/10.1007/978-3-030-84440-0_3

aim to facilitate citizens' participation in local renewable energy projects (European Commission, 2020a). These legal entities are defined as non-commercial market actors whose main goal is to provide socio-economic or environmental benefits. They can be composed of a mix of different actors, including citizens, SMEs (Small and Medium-Sized Enterprises), and local authorities, and are based on open and voluntary participation (Eur-Lex, 2020). Although the definitions of these new legal models are almost identical, renewable energy communities have a proximity requirement, which means that they are generally involved with initiatives rooted in geographically defined communities, e.g., the people of a village or the inhabitants of an island. The definition of renewable energy community, as given by the renewable energy directive, resembles the concept of CE as a community of place (Hamilton et al., 2014). However, the EU definition also highlights the participation of SMEs and local authorities, whereas CE has often been characterized by exclusively citizen-owned initiatives (e.g., *Bürgerenergie* in Germany). In light of this, we must specify that we use the term community energy (CE) only to refer to renewable energy communities that are established and run by private citizens.

The existing literature on CE is primarily based in the UK, though it also contains notable contributions from Germany, Denmark, and the Netherlands (Busch et al., 2019). As Holstenkamp (2018) noted, numerous studies have analysed CE initiatives in Central and Western Europe,

A. Isakovic
Working Group Economic Geography,
Christian-Albrechts-Universität zu Kiel, Kiel, Germany
e-mail: isakovic@geographie.uni-kiel.de

T. Hansen
Department of Food and Resource Economics (IFRO), Section for Production, Markets and Policy, University of Copenhagen, Copenhagen, Denmark

Department of Technology Management, SINTEF, Trondheim, Norway
e-mail: teis.hansen@ifro.ku.dk

and some research has been conducted in Northern Europe (e.g., Islar & Busch, 2016; Ruggiero et al., 2018; Magnusson & Palm, 2019), but there is far less extant research in Eastern European countries. Therefore, this chapter aims to address this knowledge gap by exploring CE in four Eastern European countries that have not been sufficiently studied: Estonia, Latvia, Lithuania, and Poland. We focus on these countries because they share a similar historical background (i.e., they were all governed by a communist regime and have joined the EU in recent decades). To fulfil our research goals, we rely on four case studies that were conducted as part of an EU INTERREG project (Co2mmunity https://co2mmunity.eu/). We first explore the general situation of CE in the Baltic Sea Region and identify which contextual conditions might determine the success of citizen-driven renewable energy initiatives. Subsequently, we identify (a) the drivers, (b) the barriers, (c) ways in which barriers might be overcome, and (d) project benefits.

This chapter is structured as follows: in Sect. 3.2 we summarize the literature dealing with the success factors of CE, in Sect. 3.3 we present our research material, in Sect. 3.4 we analyse and compare key contextual conditions for CE in the western and eastern portions of the Baltic Sea Region, and in Sects. 3.5 and 3.6 we present the results of our case studies. Finally, we present our conclusions in Sect. 3.7.

3.2 Overview of Community Energy Success Factors

Since the early 2000s, the literature on CE has rapidly expanded. Although it provides detailed empirical evidence of the barriers to and drivers of development of CE, most of the case-based evidence comes from a limited number of countries, in particular the UK, Germany, the Netherlands, Italy, and Denmark (Busch et al., 2021). This section presents a synthesis of this literature and provides a foundation for the subsequent consideration of success factors for CE projects in the four countries under study. We adopt an analytical perspective that gives consideration both to the role of contextual factors, policy, and project-specific factors, and to the attributes of the actors involved in the projects (Hansen & Coenen, 2015; Ruggiero et al., 2021).

3.2.1 Contextual Factors

CE projects are heavily influenced by their socio-cultural environment and by the political economy structure in which they are embedded. Socio-cultural factors, such as levels of civic engagement, educational attainment, and wealth, as well as the availability of social capital and traditions of engagement in cooperatives, associations, and other non-profit-oriented organizations, have all proven to constitute important determinants of CE development. The seminal paper by Walker and Devine-Wright (2008) and that by Radtke (2014) both highlight that civic engagement is an essential foundation for CE projects. This engagement might have its roots in a unifying cause. For example, the anti-nuclear movement in Germany, which contributed to the rise of the Green party, laid the foundation for an expansion of renewables in the country. Similarly, trust and high levels of social capital are seen as mandatory to the success of the collaborative efforts that are required to develop citizen-driven renewable energy initiatives (Walker et al., 2010). Some have argued that countries such as Denmark and Germany, which have traditions related to cooperatives, social enterprises, and engagement in work-based associations, possess favourable conditions for the development of CE (Simcock et al., 2016). Finally, communities with high levels of income and education have also been found to show a high propensity for engagement with CE projects (Radtke, 2014; Magnani & Osti, 2016).

In terms of political economy, the level of decentralization of resources and authority to sub-national levels of government can influence success. Beneficial CE programmes have been connected to high levels of municipality-level authority, as decentralization empowers local governments and allows them to take a more active role in development (Oteman et al., 2014; Gancheva et al., 2018). However, studies also show that municipalities may facilitate CE initiatives even with low levels of authority, as is seen in the Netherlands (Warbroek & Hoppe, 2017). Moreover, some municipalities with high degrees of autonomy such as those in Sweden, have been found to hinder CE development (Magnusson & Palm, 2019). Therefore, even if the decentralization of authority and resources to well-functioning sub-national governments is more markedly associated with positive effects on CE development than is decentralization in contexts where institutional capacity is limited, care must be taken when extrapolating from successful cases of decentralization (Rodríguez-Pose & Ketterer, 2019).

3.2.2 Policy

The existence of different types of policy instruments also significantly influences opportunities for the development of CE projects. In particular, energy policies, which directly influence the price of energy or provide support for small-scale energy production, are of great importance. Generally, higher energy prices incentivize investments in CE development. Consequently, policies with direct implications for the prices paid by consumers are of significant importance (Ruggiero et al., 2014, 2015). Furthermore, as highlighted by Curtin et al. (2017), multiple types of policies may provide direct financial incentives for CE, including feed-in tariffs (Fleiß et al., 2017), loan schemes (Strachan et al., 2015), and tax incentives (Mey & Diesendorf, 2018). On a more general level, the institutionalization of CE as an important component of the energy system is emphasized, as is, consequently, the belief that CE deserves further attention in energy policy (Oteman et al., 2017). Finally, policies that provide targets or visions for the future development of CE and of energy systems in general have also been found to play central roles in a considerable number of cases (for a review, see Busch et al., 2021).

3.2.3 Project-Specific Factors

The organizational characteristics of a CE project also influence its likelihood of success (Warbroek et al., 2019). Ownership structure has, in particular, been found to be of importance: co-ownership models that include local citizens significantly increase local acceptance of CE (Musall & Kuik, 2011). CE projects also profit from the early involvement of local citizens, which creates trust and eases the process of securing the necessary permits (Ruggiero et al., 2014; Walker et al., 2010). Networking with other actors, including other CE projects, broader CE initiatives, and other types of external actors, especially intermediaries, also contributes positively to CE development by providing access to different types of resources (e.g., Seyfang et al., 2013, 2014; Warbroek et al., 2019).

3.2.4 Attributes and Roles of Involved Actors

Finally, the attributes of actors directly involved in CE projects have also been found to be influential. Related to the importance of establishing networks, the leadership skills that allow CE leaders to identify, access, and exploit resources are essential for effective projects (Martiskainen, 2017). Similarly, entrepreneurial behaviour has been identified as an essential driver behind the materialization of large CE initiatives that require highly motivated pioneers to not only initiate the projects but also to create favourable policy conditions (Sperling, 2017; Young & Brans, 2017).

In terms of specific actor groups, previous studies have pointed towards key roles for intermediaries and local champions. While definitions of intermediaries abound, they are generally understood as organizations that "bridge between actors involved in situations where direct interaction is difficult due to high transaction costs (e.g., locating a suitable partner to collaborate with, disincentives to collaborate) or communication problems resulting from differences in culture, interests, and capacity to absorb or exchange knowledge" (Kivimaa et al., 2019, p. 1063). Hargreaves et al. (2013) note that intermediation creates possibilities through context-specific actions rather than through the formulation of a single blueprint from which to upscale CE solutions.

3.3 Research Material

This chapter relies on four case studies of CE projects that were conducted within an EU Baltic Sea Region INTERREG project ('Co2mmunity') that studies and facilitates the uptake of CE initiatives in the Baltic Sea Region. The four cases are located in Estonia, Latvia, Lithuania, and Poland, each of which involves distinct renewable energy technologies and actors (see Table 3.1). The cases were selected by a two-step process. In the first step, we asked a group of experts from the four countries to list possible CE projects. In the second, only cases that matched our definition of CE were included, i.e., only cases of renewable energy communities that had been established and were run by citizens. During the selection process, we followed a maximum variation sampling strategy (Patton, 1990) to achieve an appropriate degree of variation. For each case, we first collected archival data on the general conditions of CE in the Baltic Sea Region (e.g., statistics, country reports, databases and indicators, and policy documents) and then conducted semi-structured interviews with project leaders and other relevant stakeholders.

Table 3.1 List of CE cases, countries, technology used, and main actors involved

Name	Country	Type of technology	Main actors involved
Stacijas Housing Community	Latvia	Ground source heat pump	Apartment owners, building maintenance association, renewable energy company
Kagu Commercial Association	Estonia	Solar PV	Municipality of Värska, one municipal company, one NGO, private individuals
Pszczelna Street Housing Community	Poland	Solar PV	Members of the housing company, project coordinators, energy advisor
Smalininkai Community	Lithuania	Wind power	Smalininkai village community, NGOs, researchers, suppliers and contractors, local credit union

3.4 Contextual Conditions for Community Energy in the West and East Sides of the Baltic Sea Region

Figure 3.1 shows the strength of some of the key contextual factors for CE development in the Baltic Sea Region. Although in Sect. 3.2 we discussed a broader spectrum of contextual factors, in Fig. 3.1 we consider only five contextual conditions due to limitations in data availability. The strength of the five contextual conditions is measured through proxy variables that are based on numerical data collected from various sources. For civic activism and social capital, we used data from the Indices of Social Development by the International Institute of Social Studies (ISS, 2020). The civic activism indicator measures the extent to which social practices encourage active and critical engagement with political authorities. Social capital is measured in terms of the relative level of participation in voluntary activities between individuals in the same locale. For income and electricity prices, we relied on data from the statistical office of the European Union (Eurostat, 2016, 2018). In particular, we used the adjusted gross disposable income of households per capita and electricity

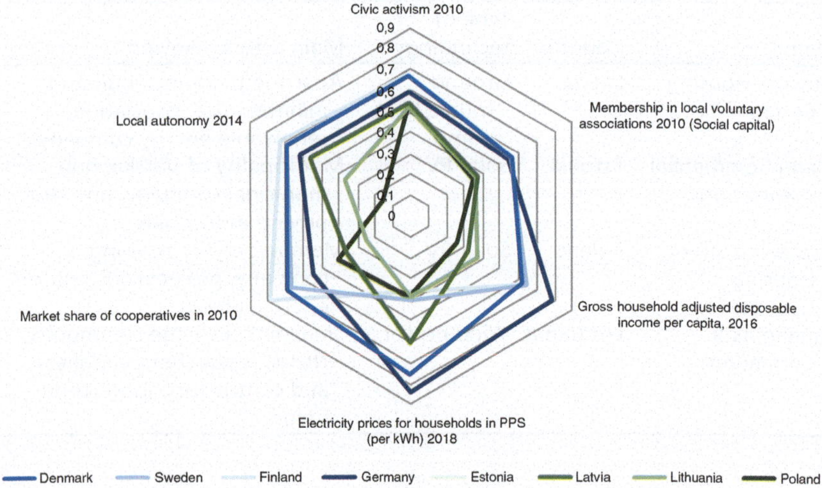

Fig. 3.1 Strength of key contextual conditions in the Baltic Sea Region

prices for households per kWh, both expressed in Purchased Power Standard (PPS). As proxy for the tradition of cooperatives, we used the market share of farmers' cooperatives, which are one of the most traditional models. Therefore, their market share can be used as a good approximation of the strength of the cooperative model in a specific country. We use data from Bijman and Iliopoulos (2014), who evaluated the market shares that farmers' cooperatives hold in different sectors and countries. For local autonomy, we used data produced by Ladner et al. (2016), who created a methodology that incorporates all the different aspects of local autonomy and used it to rank thirty-nine EU countries. We used the most recent available data for all variables; at the time this chapter was written, publication dates ranged between 2010 and 2016.

As the numerical data obtained for each of the considered contextual conditions were expressed on different scales, they were normalized (i.e., a standard score ranging from 0 to 1 was assigned in lieu of the raw scores). To normalize the data, we first calculated the mean and standard deviation of each of our variables. Then we subtracted the mean from each value and divided by the standard deviation.

Figure 3.1 shows that most of the considered variables assume smaller values (ranging between 0 and 1) for the countries on the eastern side of the Baltic Sea Region (green lines). Apart from electricity prices, Denmark, Finland, Germany, and Sweden have higher scores than Estonia, Latvia, Lithuania, and Poland. This can largely be attributed to the differing historical legacies and levels of economic development between the two sides of the region. Figure 3.1 also shows that Finland has the highest score for municipal autonomy but only a modest number of CE projects (IRENA, 2016; Ruggiero et al., 2018). Similarly, Sweden has the second-highest score for municipal autonomy, but it has far fewer CE initiatives than Denmark or Germany (IRENA, 2016). This confirms that local autonomy does not always lead to higher levels of CE development.

Table 3.2 shows historical (1991–2017) policy support for CE initiatives as well as more recent policies (beginning in 2017). We decided to take a long-term perspective on CE policy development in order to better understand how it contributed to the creation of conducive contextual conditions for CE projects. We present recent policy developments

Table 3.2 Strength of historical and current energy policies in the Baltic Sea Region

Countries	Historical policies (1991-2017)	Current policies (2017–2020)	Average
Denmark	2	1	1.5
Estonia	-2	-1	-1.5
Finland	-2	-2	-2
Germany	2	0	1
Latvia	-2	-2	-2
Lithuania	-2	-1	-1.5
Poland	-2	-2	-2
Sweden	0	0	0

5-point Likert scale: −2= very low, −1= low, 0= intermediate, 1= high, 2= very high

separately in Table 3.2 because, in most of the countries on the eastern side of the Baltic Sea Region, policies in support for CE initiatives were only recently introduced, whereas in countries in the western side of the region, policies in support of CE were concurrently reduced. The evaluation of the strength of the policy framework was conducted in two steps. In the first step, the authors evaluated the level of policy support for CE in a given Baltic Sea country by utilizing a five-level Likert scale that ranged from -2 (very low) to 2 (very high). These policy measures were retrieved from the International Energy Agency/IRENA joint policy and measures database (International Energy Agency, 2020a). In the second step, the preliminary evaluation was validated by a group of eight experts from each of the eight countries in the Baltic Sea Region. The experts were involved in a project focusing on CE in the Baltic Sea Region (Co2mmunity) and were working at energy agencies, municipalities, or other organizations that had experience with CE development. Apart from one case, the experts' evaluations were in line with the authors' assessment.

When contrasted with Fig. 3.1, Table 3.2 shows that countries like Finland and Sweden, although they would otherwise have good contextual conditions, have historically had less favourable energy policies than those of leading CE countries, such as Germany and Denmark. In addition, Finland and Sweden have very low electricity prices, which makes it even harder for CE projects to become economically viable. This highlights the fact that favourable contextual conditions are necessary but are not enough alone to stimulate CE development. Countries with the most successful CE development are those that also adopt favourable policies.

Denmark, Germany, and to some extent Sweden exhibit favourable contextual conditions and beneficial energy policies. On the other hand, Poland and Latvia must deal with both poor contextual conditions and unhelpful energy policies. Estonia and Lithuania have taken some moderate steps toward creating more conducive policy frameworks for small-scale energy production, but CE projects in those countries are challenged by unfavourable contextual factors. Finland has good contextual conditions but lacks adequate policy support. Although Fig. 3.1 and Table 3.2 usefully illustrate and help to explain the differing levels of CE development across the Baltic Sea Region, it is important to remember that these conditions are not fixed and can change over time.

3.5 Country Reviews

After reviewing some of the key contextual conditions for CE development throughout the Baltic Sea Region, this section can now focus on the current status of CE in the eastern portion of the region. Based on the review of policy support illustrated in the previous section (Table 3.2), we divided the countries into two groups: Lithuania and Estonia, which have adopted a modest policy mix to promote CE initiatives (mainly in the case of housing communities), and Latvia and Poland that are still lagging behind.

3.5.1 Lithuania and Estonia: Modest Progress in the Right Direction

Lithuania is heavily dependent on imported electricity that is mainly from Sweden and Poland. This is partly due to the shutdown of the country's only nuclear reactor as a circumstance of Lithuania's accession to the EU. In its latest energy strategy, Lithuania has set a goal to satisfy 70 per cent of the country's electricity demand with domestic sources (up from the current 30 per cent). In addition, the share of renewables in power generation is projected to increase from 23 per cent to 45 per cent (European Commission, 2020b). These goals are set for the year 2030; even though very few CE projects currently exist in the country, CE is expected to eventually play a role. Consequently, the role of CE is recognized in the government's energy independence strategy (Government of Lithuania, 2018).

Lithuania has taken the first steps needed to encourage a larger diffusion of energy prosumerism. According to government targets, energy prosumers (citizens that produce and consume energy) will account for at least 30 per cent of all consumers by 2030 and at least 50 per cent by 2050 (Government of Lithuania, 2018). Besides setting a clear target for energy prosumers, Lithuania has also introduced a net metering scheme to support residential and commercial electricity production by solar PV. Despite recent policy efforts, CE initiatives in Lithuania face numerous barriers. Information on technical solutions and legal forms is often hard to access, and low income levels paired with a general reluctance by banks to provide favourable loans often lead to financial challenges. Finally, the country's Soviet heritage creates a further cultural barrier. The repression of an independent civil society by the Soviet regime has led to a situation in which civil activism in Lithuania and other Eastern European states is lower than in their Western counterparts (Aidukaitė, 2013). In addition, the oftentimes negative experience of forced collectivization of farms has created a level of distrust towards cooperatives in general, including CE initiatives. Despite these hurdles, CE can play an important role in reaching the country's self-sufficiency goals if further supportive policies are issued.

Estonia is in a similar situation to Lithuania. While some progress is visible and some favourable policies are in place, CE is generally still in early stages throughout the country. With more than 80 per cent of the overall share, fossil sources remain the main source of energy (International Energy Agency, 2020b). The lion's share of these fossil fuels comes from the country's domestic reserves of oil shale, a rock formation that contains kerogen (organic matter that can be turned into fuel). This makes the Estonian economy even more carbon-centric than that of Lithuania (International Energy Agency, 2019). Though wind and biomass have increased the overall share of renewable energy in the Estonian electricity mix since 2008, CE still faces cultural barriers similar to those in Lithuania. Collective action and ownership arrangements are still viewed with suspicion because of the country's Soviet heritage. Further barriers include low electricity prices, unfavourable interest rates for loans related to CE projects, and a general lack of information. In recent years, however, some favourable policy instruments have been introduced. For instance, small power plants of up to 50 kW are now eligible for a feed-in tariff. This policy is complemented by an energy efficiency fund for old buildings that provides subsidies of up to 30 per cent for renovation projects focusing on energy self-consumption in apartment buildings. This policy mix grants hope for CE growth in Estonia's future.

3.5.2 Poland and Latvia: Still Lagging Behind

Whereas Lithuania and Estonia have shown some progress in the establishment of CE projects, the situation in Poland and Latvia is not as positive. Similar to the other countries in the study, Poland's electricity sector is dominated by state-owned incumbents from the fossil fuel industry. The country still relies heavily on coal, which is a clear barrier for renewable energy projects. Political support for RE is sparse, legislation for cooperatives is outdated, and the legal framework for CE is characterized by instability. Further, regulatory barriers make it difficult for CE initiatives to find niches in which to establish projects. For example, it is not possible to set up a project in Poland in which individual households in apartment blocks can fully benefit from energy self-consumption. Again, the Soviet heritage imposes further cultural barriers on Poland, where cooperative arrangements are still negatively associated with state socialism (Beckmann et al., 2015). Another barrier is the general financial condition in Poland; citizens often lack the capital to make investments in CE projects, and loans are not readily available. Last, unfavourable distancing rules make further expansion of modern wind energy nearly impossible in Poland. The Distancing Act of 2016 set a minimum distance of ten times the height of the tip of the blade for large turbines in any residential or mix-purpose building or in an area of environmental value (Skrzypczyk, 2020).

Not surprisingly, there are very few examples of CE projects in Poland. However, some supporting policies are directed towards *energy clusters*. These clusters are related to the CE model but differ in two crucial points. First, energy clusters do not exclusively focus on RE sources but include all forms of local energy sources. Second, they are not limited to citizens but are open to private companies. This focus on multiple actors qualifies them more as renewable energy communities than as CE projects. As a result of the adopted definition, the driving force behind the establishment of energy clusters in Poland are local authorities and businesses with a very low level of citizen participation. However, despite the limitations of the energy cluster model, it can serve as an opportunity for CE groups and can be used to phase out coal use in the long-term future.

Like Lithuania, one of the cornerstones of Latvia's energy policy has been a goal to reduce dependence on imported energy. After the fall of the Soviet Union, imported energy accounted for almost 90 per cent of Latvia's total consumption. Over the years, increased use of renewable energy, especially biomass, has enhanced the country's energy efficiency and increased its energy independence. However, Latvia still relies heavily on imported fossil energy (OECD, 2019). The country's renewable energy policy has been inconsistent and is often hampered by incumbents' vested interests and political instability. There are currently very few CE projects in Latvia. Nevertheless, there are a handful of hopeful exceptions, including initiatives carried out by municipalities to promote increased supplies of biomass from local cooperatives and examples of renewable energy production in housing communities. As in the countries previously examined, one of the main barriers for CE in Latvia is general distrust of collective ownership that is a consequence of the country's Soviet heritage.

3.6 Drivers, Barriers, and Benefits of Community Energy Projects

Table 3.3 offers an overview of the main drivers, barriers, and benefits of CE projects that emerged from the interviews we conducted. We found that favourable preconditions contributed to the success of the project and that national and regional visions or plans proved to be important factors in the success of initiatives. For instance, this type of planning served as a favourable precondition for CE projects in Estonia and Lithuania. Other preconditions included a general sense of community to which actors could appeal and suitable ownership models for CE in the legal framework of the country. Finally, municipal networks can play a supporting role in local development and cooperation.

Our case studies unveiled various drivers behind the development of CE projects. First, financial instruments and key policy interventions, like loans for citizen groups at preferential rates, direct subsidies for RE projects, or feed-in tariffs, can promote CE. Second, these projects are often designed to address the need for local economic development; many rural regions in the eastern part of the Baltic Sea Region have experienced relocation of jobs and, consequently, the migration of young people to urban areas. In accordance with previous research (e.g., Berka & Creamer, 2018; Brummer, 2018; Busch & McCormick, 2014), our case studies confirm that CE projects can create new jobs, thereby increasing tax income for municipalities and investment in local infrastructure. Third, CE projects were often motivated by the prospect of reducing energy costs. Although the literature on CE has often described idealism as one of the main drivers, in our cases idealism was not the main motivator. Although some interviewees reported that they wanted to set a good example with their project, this was considered an extra benefit, not the reason for initial involvement. In addition, climate change and other environmental factors were not mentioned at all.

In line with previous research (Ivask et al., 2021), we found that CE projects face a number of barriers in these countries, the first of which are economic barriers. Low energy prices on the national electricity markets can act as a barrier to the economic viability of projects. Fixed feed-in

Table 3.3 Summary of findings from the case studies

Case	Favourable preconditions	Drivers	Barriers	Actions to overcome barriers	Project benefits
Stacijas Housing Community (Latvia)	EU support for energy efficiency	Installation grant, Reduce energy costs, Be a good example	Low citizen awareness of renewable energy, Lack of national support mechanisms, High upfront investment, Low financial leverage of citizens, No loan guarantees available	Building maintenance company focused their efforts on seeking and applying for EU grant funding	Improved indoor temperature, Elimination of mould and lower humidity levels, Educational effect, Reduced energy costs
Kagu Commercial Association (Estonia)	Sense of community, Network of municipalities promoting development and cooperation, Regional plan promoting energy self-sufficiency, Suitable ownership model	Depopulation, Local economic development, Reduce energy costs, Feed-in tariff, Be a good example	Energy regulation, Grants only for on-site energy consumption, Low feed-in tariff, Costs of grid connection	Find a company that could buy the generated electricity, Situating the PV installation on the company's property	There are not yet any benefits because the project is still in the planning phase

Pszczelna Street Housing Community (Poland)	Local support for energy self-consumption	Installation grant, Preferential loan conditions, Reduce energy costs	Low citizen awareness of renewable energy, Lack of national support mechanisms, Bureaucracy	Project leaders focused their efforts on seeking and applying for grant funding	Reduced energy costs, Increased apartment value, Educational effect, Surplus reinvested into renovation fund
Smalininkai Community (Lithuania)	State's ambition to increase wind power, Wind power projects in the region	Economic development	Lack of experience, High upfront investment, High-interest loan, Community was doubtful	Intense communication and high degree of transparency	Failed project but precious learning experience

tariffs have been proposed as a solution to this problem; however, if these are set too low, they can also hinder CE development. Further economic problems can occur when CE groups are faced with substantial fees for grid connection. Finally, the high costs of RE investment are a barrier for some community groups. CE projects require a substantial sum to cover the upfront investment. These costs are often paired with high interest rates for loans, which together can be too much for a community to bear.

Apart from the economic barriers, regulation is often not favourable in the eastern Baltic Sea Region. Historically, regulation in the region has been built around the needs of big incumbents from the fossil fuel sector and, therefore, no adequate regulation exists for niche initiatives such as CE projects. When exceptions do exist, support mechanisms are often limited to a narrow scope of possibilities, e.g., energy renovations or self-consumption. This arrangement conflicts with the needs of local communities to design and implement projects in response to local needs and potentials. A further barrier arises from the widespread mistrust in the support mechanisms provided by the state. This leads to citizens being hesitant to risk their private funding, as they fear that a change in subsidies or feed-in tariffs might jeopardize their investment in a CE project. For example, Lithuania and Estonia both have a national strategy and support mechanisms for small-scale RE generation, but they have not led to the hoped-for breakthrough for CE projects.

The last set of barriers arises from a lack of knowledge and from low levels of general awareness of the benefits of RE technology, which seems to be more widespread in this part of the Baltic Sea Region than in the western section. State and intermediary actors have also not been successful in communicating the advantages of CE projects to citizens. Finally, a general lack of experience with community-led energy initiatives means that there are no previous examples upon which to build.

Key actors play a crucial role in overcoming the numerous barriers. Oftentimes key actors are local champions who are in one or another way involved in the CE project. These can, for example, be local mayors (Busch & McCormick, 2014) or simply engaged citizens in a housing association (Ruggiero et al., 2021). Key actors are often not individuals but rather small groups of people (Warbroek et al., 2019). In line with previous research (Ruggiero et al., 2021), we found that local champions may attempt to make use of favourable contextual conditions and drivers to address and overcome barriers. We identified three strategies that local actors adopted to overcome these obstacles to development. The first focuses upon intensive community engagement, the second on external sources of financing, and the third on finding partners that can assist with restrictive regulation. Community engagement was helpful in overcoming doubt in two of the cases we investigated. In both Estonia and Lithuania, proponents of the CE project used this approach to shift opinion in the community and gain wide support. In the case of Smalininkai village in Lithuania, intensive communication and a high degree of transparency regarding the related economic situation were needed in order to address prevalent doubts concerning the project.

Funding is a central challenge for all CE projects. The Polish case, the Pszczelna Street Housing Community, is no exception to this; the project managers diversified their fundraising strategy, tapping into several funding streams in order to raise the required funds. As Estonian energy regulation does not allow for energy prosumer models in which the place of production and consumption are not the same, the key actors in the Kagu project in Estonia had to find a creative way to overcome regulatory barriers. The solution was to establish a partnership with a local company. The CE group and the company struck a power purchase partnership agreement which stipulated that all of the generated energy would be bought by the partner company.

CE projects can produce a host of benefits for local communities. In most of our cases, the project benefits included savings in energy costs. Another benefit was the increase in apartment value once a new renewable energy system was installed. Further benefits included enhanced educational opportunities, improved indoor climate, and positive effects on the local economy through indirect jobs, local tax revenue for municipalities, and educational effects.

3.7 Conclusions

The objective of this chapter was to explore CE in a new geographical context, the eastern side of the Baltic Sea Region, and to contrast it with its western counterpart. Building on four case studies, we studied the drivers, barriers, and benefits of CE initiatives in Estonia, Latvia, Lithuania, and Poland. The analysis of these cases shows that CE initiatives are, in general, still rare in the eastern Baltic Sea Region; the analysed countries have far fewer CE initiatives than do the countries on the western side of Europe. This is in line with the findings of a recent report by the Joint Research Centre of the European Commission on energy communities in Europe (Caramizaru & Uihlein, 2020). The substantial difference in the number of CE initiatives can be attributed to both contextual conditions and energy policies. Additionally, the actors involved in the studied initiatives often exploited favourable preconditions, such as the presence of national or local planning. In general, CE projects were mainly driven by ad hoc EU or regional grant funding, possibly to reduce energy costs and promote local economic development. Projects face numerous hurdles, mainly distrust and lack of interest towards community ownership and collective methods of energy production. This is due to the historical background of this part of the Baltic Sea Region, which operated under a socialist regime for several decades. Other barriers include low awareness of renewable energy sources, low income, inadequate regulation, and lack of effective finance mechanisms for community-led projects. In addition to direct benefits, such as reduced energy costs and improved living conditions, the CE projects in this study provided indirect benefits to the local economy in terms of jobs and tax revenues for the municipalities.

To stimulate citizens' renewable energy projects, some countries on the eastern side of the Baltic Sea Region have taken small steps to promote a CE approach. The recent EU renewable energy and electricity directives have paved the way for further progress, although significant changes may not immediately emerge due to the generally slow pace of national legislative frameworks. Moreover, as the transposition process of EU rules into national legislation leaves significant room for specific provisions determined by individual member states, the effectiveness of any national measures remains to be seen. However, by and large, the new EU directives will certainly contribute to the creation of more favourable conditions for CE in the eastern Baltic Sea Region and will remove some of the primary obstacles that might hinder, for example, energy self-consumption in apartment buildings. Moreover, they will force countries in this region to acknowledge the role of citizens and civil society organizations in the transition towards a low carbon society. However, the degree of citizen participation in local renewable energy initiatives will not only depend on the introduction of favourable legislation but also, to a large extent, on efforts to overcome cultural barriers.

References

Aidukaitė. (2013). Community Mobilizations Around Housing and Local Environment: Insights into the Case of Vilnius. Sociology. *Thought and Action, 32*(1), 136–151.

Beckmann, V., Otto, I. M., & Tan, T. (2015). Overcoming the Legacy of the Past? Analyzing the Modes of Governance Used by the Polish Agricultural Producer Groups. *Agricultural Economics – Czech, 61*, 222–233.

Berka, A. L., & Creamer, E. (2018). Taking Stock of the Local Impacts of Community Owned Renewable Energy: A Review and Research Agenda. *Renewable and Sustainable Energy Reviews, 82*, 3400–3419.

Bijman, J., & Iliopoulos, C. (2014). Farmers' Cooperatives in the EU: Policies, Strategies, and Organization. *Annals of Public and Cooperative Economics, 85*, 497–508.

Brummer, V. (2018). Community Energy – Benefits and Barriers: A Comparative Literature Review of Community Energy in the UK, Germany and the USA, the Benefits It Provides for Society and the Barriers It Faces. *Renewable and Sustainable Energy Reviews, 94*, 187–196.

Busch, H., & McCormick, K. (2014). Local Power: Exploring the Motivations of Mayors and Key Success Factors for Local Municipalities to Go 100% Renewable Energy. *Energy, Sustainability and Society, 4*(5). https://doi.org/10.1186/2192-0567-4-5

Busch, H., Ruggiero, S., Isakovic, A., Faller, F., & Hansen, T. (2019). *Co2mmunity Working Paper No. 2.1: Scientific Review Paper on CE Drivers and Barriers*. Available at: http://co2mmunity.eu/outputs/download-area. Accessed 6 Mar 2019.

Busch, H., Ruggiero, S., Isakovic, A., & Hansen, T. (2021). Policy Challenges to Community Energy in the EU: A Systematic Review of the Scientific Literature. *Renewable and Sustainable Energy Reviews, 151*, 111535.

Caramizaru, A., & Uihlein, A. (2020). *Energy Communities: An Overview of Energy and Social Innovation*. EUR 30083 EN, Publications Office of the European Union, Luxembourg, ISBN 978-92-76-10713-2. https://doi.org/10.2760/180576, JRC11943.

Curtin, J., McInerney, C., Gallachóir, Ó., & B. (2017). Financial Incentives to Mobilise Local Citizens as Investors in Low-Carbon Technologies: A Systematic Literature Review. *Renewable and Sustainable Energy Reviews, 75*, 534–547.

EUR-Lex. (2020). *Proposal for a Directive of the European Parliament and of the Council on the Promotion of the Use of Energy from Renewable Sources (Recast)*. Available at: https://eurlex.europa.eu/legalcontent/EN/TXT/?uri=CELEX%3A52016PC0767. Accessed 23 Feb 2020.

European Commission. (2020a). *Clean Energy for All Europeans Package*. Available at: https://ec.europa.eu/energy/topics/energy-strategy/clean-energy-all-europeans_en. Accessed 3 Apr 2020.

European Commission. (2020b). *Lithuania: Summary of the Commission Assessment of the Draft National Energy and Climate Plan 2021–2030*. Available at: https://ec.europa.eu/energy/sites/ener/files/documents/necp_factsheet_lt_final.pdf. Accessed 22 Dec 2020.

Eurostat. (2016). *Adjusted Gross Disposable Income of Households Per Capita in PPS*. Available at: https://data.europa.eu/euodp/en/data/dataset/88K43em2njs6X2S7qSVA. Accessed 21 Dec 2020.

Eurostat. (2018). *Electricity Prices for Households and Industrial Users (PPS)*. Available at: https://www.pordata.pt/en/MicroPage.aspx?DatabaseName=Europe&MicroName=Electricity+prices+for+households+and+industrial+users+(PPS)&MicroURL=1479&. Accessed 21 Dec 2020.

Fleiß, E., Hatzl, S., Seebauer, S., & Posch, A. (2017). Money, Not Morale: The Impact of Desires and Beliefs on Private Investment in Photovoltaic Citizen Participation Initiatives. *Journal of Cleaner Production, 141*, 920–927.

Gancheva, M., O'Brien, S., Crook, N., & Monteiro, C. (2018). *Models of Local Energy Ownership and the Role of Local Energy Communities in Energy Transition in Europe*. European Union and the Committee of the Regions. Available at: https://publications.europa.eu/en/publication-detail/-/

publication/667d5014-c2ce-11e8-9424-01aa75ed71a1/language-en/format-PDF. Accessed 5 Mar 2019.

Government of Lithuania. (2018). *National Energy Independence Strategy (Unofficial Translation)*. Available at: https://enmin.lrv.lt/uploads/enmin/documents/files/National_energy_independence_strategy_2018.pdf. Accessed 21 Dec 2020.

Hamilton, J., Mayne, R., Parag, Y., & Bergman, N. (2014). Scaling Up Local Carbon Action: The Role of Partnerships, Networks and Policy. *Carbon Management, 4*, 463–476.

Hansen, T., & Coenen, L. (2015). The Geography of Sustainability Transitions: Review, Synthesis and Reflections on an Emergent Research Field. *Environmental Innovation and Societal Transitions, 17*, 92–109.

Hargreaves, T., Hielscher, S., Seyfang, G., & Smith, A. (2013). Grassroots Innovations in Community Energy: The Role of Intermediaries in Niche Development. *Global Environmental Change, 23*, 868–880.

Holstenkamp, L. (2018). Einleitende Anmerkungen zum Ländervergleich: Definition von Bürgerenergie, Länderauswahl und Überblick über Fördermechanismen. In L. Holstenkamp & J. Radtke (Eds.), *Handbuch Energiewende und Partizipation*. Springer VS.

International Energy Agency. (2019). *Estonia 2019 Review*. Available at: https://webstore.iea.org/energy-policies-of-iea-countries-estonia-2019-review. Accessed 17 Apr 2020.

International Energy Agency. (2020a). *Policies Database*. Available at: https://www.iea.org/policies?source=IEA%2FIRENA%20Renewables%20Policies%20Database. Accessed 21 Dec 2020.

International Energy Agency. (2020b). *Estonia—Key Energy Statistics 2018*. Available at: https://www.iea.org/countries/estonia. Accessed 16 Mar 2020.

IRENA. (2016). *Renewable Energy Cooperatives in Europe*. Available at: https://www.irena.org/-/media/Files/IRENA/Agency/Events/2016/Jan/18/ENERCOOP-People-Power%2D%2DRenewable-Energy-Cooperatives-in-Europe.pdf?la=en&hash=474259287FCBD81FC4738041B3B87FA22592E356. Accessed 3 Apr 2020.

Islar, M., & Busch, H. (2016). We Are Not in This to Save the Polar Bears!—The Link Between Community Renewable Energy Development and Ecological Citizenship. *Innovation: The European Journal of Social Science Research, 29*(3), 303–319.

ISS. (2020). *Data Access*. Available at: https://isd.iss.nl/data-access/. Accessed 21 Dec 2020.

Ivask, N., Kärbo, N., Kikas, M., Vabamägi, A., Wagner-Vogel, G., Ståhl, P., Ruggiero, S., Makowska, J., Marczak, A., Cederqvist, A., Kamenders, A., Francis, I., Tuisku, P., Viljanen, A., Murauskaite, L., Johannsen, M., & Keller, P. (2021). *Policy Recomendations for Renewable Energy Communities Development*. Aalto Working Papers. https://aaltodoc.aalto.fi/bitstream/handle/123456789/107003/isbn9789526403359.pdf?sequence=1. Accessed 21 Mar 2021.

Kivimaa, P., Boon, W., Hyysalo, S., & Klerkx, L. (2019). Towards a Typology of Intermediaries in Sustainability Transitions: A Systematic Review and a Research Agenda. *Research Policy, 48*, 1062–1075.

Ladner, A., Keuffer, N., & Baldersheim, H. (2016). Measuring Local Autonomy in 39 Countries (1990–2014). *Regional & Federal Studies, 26*, 321–357.

Magnani, N., & Osti, G. (2016). Does Civil Society Matter? Challenges and Strategies of Grassroots Initiatives in Italy's Energy Transition. *Energy Research & Social Science, 13*, 148–157.

Magnusson, D., & Palm, J. (2019). Come Together—The Development of Swedish Energy Communities. *Sustainability, 11*(4), 1056.

Martiskainen, M. (2017). The Role of Community Leadership in the Development of Grassroots Innovations. *Environmental Innovation and Societal Transitions, 22*, 78–89.

Mey, F., & Diesendorf, M. (2018). Who Owns an Energy Transition? Strategic Action Fields and Community Wind Energy in Denmark. *Energy Research & Social Science, 35*, 108–117.

Musall, F. D., & Kuik, O. (2011). Local Acceptance of Renewable Energy—A Case Study from Southeast Germany. *Energy Policy, 39*, 3252–3260.

OECD. (2019). *Fossil Fuel Support Country Note: Latvia*. Available at: https://www.google.com/url?sa=t&rct=j&q=&esrc=s&source=web&cd=3&ved=2ahUKEwjF7OX38e7oAhUHxosKHaFMBPIQFjACegQIAxAB&url=http%3A%2F%2Fstats.oecd.org%2Fwbos%2Ffileview2.aspx%3FIDFile%3De9fa2e7c-4b74-47b1-b493-0ced1fd5df3e&usg=AOvVaw0kyXH8tYUJBQjEdabz8qGo. Accessed 17 Apr 2020.

Oteman, M., Kooij, H.-J., & Wiering, M. A. (2017). Pioneering Renewable Energy in an Economic Energy Policy System: The History and Development of Dutch Grassroots Initiatives. *Sustainability, 9*, 550.

Oteman, M., Wiering, M., & Helderman, J. K. (2014). The Institutional Space of Community Initiatives for Renewable Energy: A Comparative Case Study of the Netherlands, Germany and Denmark. *Energy, Sustainability and Society, 4*, 11.

Patton, M. Q. (1990). *Qualitative Evaluation and Research Methods* (2nd ed.). Sage.

Radtke, J. (2014). A Closer Look Inside Collaborative Action: Civic Engagement and Participation in Community Energy Initiatives. *People Place and Policy Online, 8*, 235–248.

Rodríguez-Pose, A., & Ketterer, T. (2019). Institutional Change and the Development of Lagging Regions in Europe. *Regional Studies*, 1–13.

Ruggiero, S., Busch, H., Hansen, T., & Isakovic, A. (2021). Context and Agency in Urban Community Energy Initiatives: An Analysis of Six Case Studies from the Baltic Sea Region. *Energy Policy, 148*, Part A.

Ruggiero, S., Martiskainen, M., & Onkila, T. (2018). Understanding the Scaling-Up of Community Energy Niches Through Strategic Niche Management Theory: Insights from Finland. *Journal of Cleaner Production, 170*, 581–590.

Ruggiero, S., Onkila, T., & Kuittinen, V. (2014). Realizing the Social Acceptance of Community Renewable Energy: A Process-Outcome Analysis of Stakeholder Influence. *Energy Research & Social Science, 4*, 53–63.

Ruggiero, S., Varho, V., & Rikkonen, P. (2015). Transition to Distributed Energy Generation in Finland: Prospects and Barriers. *Energy Policy, 86*, 433–443.

Seyfang, G., Hielscher, S., Hargreaves, T., Martiskainen, M., & Smith, A. (2014). A Grassroots Sustainable Energy Niche? Reflections on Community Energy in the UK. *Environmental Innovation and Societal Transitions, 13*, 21–44.

Seyfang, G., Park, J. J., & Smith, A. (2013). A thousand Flowers Blooming? An Examination of Community Energy in the UK. *Energy Policy, 61*, 977–989.

Simcock, N., Willis, R., & Capener, P. (2016). *Cultures of Community Energy—International Case Studies*. British Academy. Available at: https://www.thebritishacademy.ac.uk/sites/default/files/CoCE_Policy%20Report%20_%20online_0.pdf. Accessed 12 May 2018.

Skrzypczyk, A. (2020). *Who Doesn't Like Polish Wind Power?* Available at: https://energytransition.org/2020/06/22417/. Accessed 22 Dec 2020.

Sperling, K. (2017). How Does a Pioneer Community Energy Project Succeed in Practice? The Case of the Samsø Renewable Energy Island. *Renewable and Sustainable Energy Reviews, 71*, 884–897.

Strachan, P. A., Cowell, R., Ellis, G., Sherry-Brennan, F., & Toke, D. (2015). Promoting Community Renewable Energy in a Corporate Energy World. *Sustainable Development, 23*, 96–109.

Walker, G., & Devine-Wright, P. (2008). Community Renewable Energy: What Should It Mean? *Energy Policy, 36*, 497–500.

Walker, G., Devine-Wright, P., Hunter, S., High, H., & Evans, B. (2010). Trust and Community: Exploring the Meanings, Contexts and Dynamics of Community Renewable Energy. *Energy Policy, 38*, 2655–2663.

Warbroek, B., & Hoppe, T. (2017). Modes of Governing and Policy of Local and Regional Governments Supporting Local Low-Carbon Energy Initiatives; Exploring the Cases of the Dutch Regions of Overijssel and Fryslân. *Sustainability, 9*(1), 75.

Warbroek, B., Hoppe, T., Bressers, H., & Coenen, F. (2019). Testing the Social, Organizational, and Governance Factors for Success in Local Low Carbon Energy Initiatives. *Energy Research & Social Science, 58*, 101269.

Young, J., & Brans, M. (2017). Analysis of Factors Affecting a Shift in a Local Energy System Towards 100% Renewable Energy Community. *Journal of Cleaner Production, 169*, 117–124.

4

Clean Energy Transition in Southeast Europe: The Paradigm of Greece from a Fossil Fuel Mediator to a Community Energy Hub

Nikolaos Ziozas and Theocharis Tsoutsos

4.1 Introduction

Despite the intense emergence of local energy projects in the European continent after the 1990s (RESCOOP plus, 2020), the concept of energy community (EnC) in Greece used to be until recently, a relatively innovative idea from both civil society and political perspectives. The meaning of the term "community energy" is not used in public, academic or political debates; whereas a more commercialized and generalized term, that of "*ενεργειακές κοινότητες*" (energy communities) is used to describe community-based energy projects initiated by local authorities, municipalities, or small groups of individuals. There was a limited number of references to EnC projects in published researches, grey literature and online data; a condition, which at the same time places the current

N. Ziozas (✉) • T. Tsoutsos
School of Chemical & Environmental Engineering, Renewable and Sustainable Energy Lab, Technical University of Crete, Chania, Greece
e-mail: nziozas@enveng.tuc.gr; theocharis.tsoutsos@enveng.tuc.gr

© The Author(s), under exclusive license to Springer Nature Switzerland AG 2021

F. H. J. M. Coenen, T. Hoppe (eds.), *Renewable Energy Communities and the Low Carbon Energy Transition in Europe*, https://doi.org/10.1007/978-3-030-84440-0_4

chapter as a challenge for an in-depth examination of Greek EnC based on close cooperation with Greek EnCs.

Here we approach the Greek EnC as a creative space of community-led energy initiatives with the potential to scale up and destabilize the fossil fuel regime in Greece. For this purpose, we employ the approach of the Strategic Niche Management (SNM) framework as a guidance tool for the examination of internal dynamics and processes of Greek EnCs through the lens of Greek REScoops' activities. The following sections include:

- the emergence of EnC from the Greek legislative viewpoint;
- Greek Community Energy as a socio-technical niche as a case study using the SNM approach;
- three case studies representing different typologies, as lighthouses;
- lessons learned from *in situ* visits;
- recommendations for fruitful creation and operation of EnCs.

4.2 The Emergence of Community Energy in Greece

Although hundreds of energy cooperatives (EnCs) were activated in western Europe several years ago, providing their members with numerous benefits under different local/national schemes (Sifakis et al., 2019, 2020), in Greece, there were limited numbers of community-led energy initiatives.

However, two innovative historical cases are notable. Specifically, Kithnos island, after an initiative of the state-owned Public Power Corporation, created in 1982 the first 100 kW hybrid energy system in the EU, combining a wind/PV power system with battery storage (Various, 2002). Respectively, the case study of Anavra municipality constituted one of the most successful energy initiatives at a community level. The community of Anavra based on a robust sustainable plan of the local authorities in 1990 and supported by the majority of its citizens, managed to create three livestock parks as well as the construction of a 17.5 MW wind farm with the capacity of twenty wind turbines, enhancing the energy sustainability of the community and the active participation of its civil society (Ziozas, 2019).

Even though the identification of the barriers that prevent Greek EnC initiatives to emerge and scale-up is part of the EnC concept, we will try to emphasize a landscape of challenges that decelerates the emergence of EnC in Greece. From this viewpoint, the literature review focuses on two fundamental dimensions of Greek EnC; the community development in the prism of political and legislative independence from the State and the diffusion of technologies exploiting renewable energy sources (RES) in the national energy mix from an energy policy perspective.

First, the absence of EnC initiatives from the Greek "energy map" can be easily explained by the enormous gap between local communities and RES projects (Tampakis et al., 2017). This is a result of the limited governmental and economic freedoms that municipalities and communities hold in Greece due to the centralized character of the State (Oecd.org, 2019). The first attempt for a more efficient subnational governance occurred in 2010 with the introduction of Law 3852/2010 ("New architecture of self-governance and decentralized administration"), or in short, with the Kallikratis reform (Law 3852/2010). Based on this reform, 1033 municipalities and communities were replaced by 325 municipalities into a two-level governmental structure (Various, 2002). The decentralized purpose of the Kallikratis reform is expressed by the transfer of power and responsibilities to local governments and their alignment with EU funds.

The second attempt of the government to boost the local government and economy materialized with the Cleisthenes reform (Law 4555/2018). This reform reinforced further the responsibilities of local governments and enhanced the local democratic electoral system. On the other hand, in the RES sector, this centralized status quo has a crucial impact on the development of local energy projects. In parallel, the limited local economic resources, the anelastic bureaucratic procedures, and the lack any beneficial links between communities and RES initiatives shaped a clear private business model in the relations between the community and RES. Indicatively, only 16.4 per cent of the subnational government expenditures are addressed to the local environmental protection. At the same time, the compensatory benefits from RES projects for local communities correspond only to 3 per cent of the project's revenues. Further challenges to EnC constitute the limited available grid capacity for RES projects, the low levels of social acceptance, and weak civil society activity. Therefore, the concept of National EnCs was represented by a limited number of isolated local energy projects, mainly pilots and experimental integration of RES novelties into communities.

Law 4513/2018 for energy communities tried to set a promising and fruitful legislative framework for the activation of the EnC concept in Greece (Greek Law 4513/2018). The legislation managed to define and shield the role of EnCs in the energy sector by aligning the active engagement of citizens to the development of RES projects and technologies (Gancheva et al., 2018). The identification of EnCs as legal entities fuelled the interest and the environmental consensus of a wide range of civil society groups (e.g., cooperatives, individuals, volunteers, public authorities, private firms), achieving until present a number of 319 registered EnCs in the general commercial assembly (GEMI) (Businessregistry.gr, 2020). Although a limited number of them are currently operating, EnC initiatives have also been noticed with different legal or business forms. Among them, they have placed the present examined cases of study, namely the Sifnos Island Cooperative, the Electra Energy Cooperative and the Energy Cooperative of Karditsa. It is exceptional that these three cooperatives are members of the REScoop.eu federation and constitute one of the few active energy cooperatives in the field of EnC. Therefore, it is essential to decode the Greek EnC approach through the lens of Greek REScoops and their internal dynamic processes.

A very recent Law 4759/2020, obliges all EnCs from 2022 to compete with private investors in bids to ensure the operational support of RES projects (Law 4759/2020). Given that it is practically impossible to compete openly, this regulation virtually eliminates potential incentives to set up an EnC from citizens and local authorities in the short term. According to Greenpeace and RESCoops, "It opposes the European institutional framework which identifies the specific characteristics of EU Communities, recognizes their social and developmental benefits and demands the protection of the inalienable right of citizens to participate in energy markets" (Rescoop.eu et al., 2021).

4.3 Greek Community Energy as a Socio-Technical Niche: A Strategic Niche Management Approach

The co-evolution of our ongoing fossil fuels-based energy system with our society domains over the years posed the challenge for broad systemic research in order to follow a sustainable energy shift, instead of a unilateral technological examination of the problem (Verbong & Loorbach, 2012). To capture the dynamics of this system-wide transformation, scholars from socio-technical innovation studies have conceptualized the role of niches as shielding spaces where novel socio-technical configurations and practices flourish through experimentation, development and can potentially overthrow unsustainable regimes (Smith et al., 2015; Smit & Raven, 2012). Building upon the momentum of Strategic Niche Management (SNM) literature on niche theories, scholars have researched a diverse set of fields (i.e., energy, organic and local food systems, eco-housing, and complementary currencies) as socio-technical niches, where radical and sustainable community-driven innovations emerge (Seyfang et al., 2014) Among them, a flourishing sustainable grassroots niche, which is a central concept of the present research, constitutes the EnC niche.

A first glance at the status quo of EnC in Greece through the SNM perspective could be provided by the conceptualization of niches development phases according to Fig. 4.1. Specifically, Geels and Deuten conceptualized the development of shared technological knowledge through

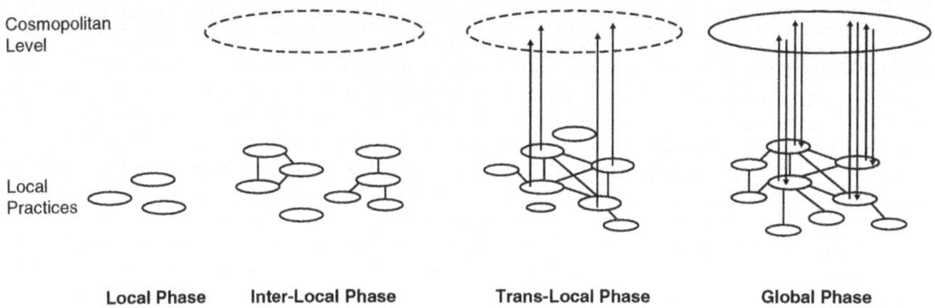

Fig. 4.1 Phases of the development of shared technological knowledge. (*Source*: Geels & Deuten, 2006)

a non-linear pathway from an emerging group of local projects to a "cosmopolitan" niche level (Geels & Deuten, 2006). Apart from the local-to-global niche conceptualization, early SNM literature has defined the niche development process central to the concept of niche nurturing, identifying three fundamental criteria for its evaluation (Schot & Geels, 2008); the so-called internal niche processes of articulation of expectations and visions, learning and building of social networks.

SNM scholars point out that that niche-building could be successful and influential if **expectations and visions** are:

- tangible (realistic targets and high-quality context);
- robust (broad acceptance and common support), and
- specific in terms of structural design and timetable

Social networks are broad (including a diverse spectrum of stakeholders) and deep by possessing a solid structure and actors that can materialize the promised expectations through internal resources and efforts, and *learning* processes overcome the traditional direction of accumulation of data and facts (e.g., first-order learning), and foster shifts in the cognitive frames and assumptions (e.g., second-order learning). Consequently, socio-technical niches become influential to the extent that all the processes mentioned above are coherent and robust to trigger a broader systemic change (Geels, 2002; Geels & Raven, 2006). Thus, our research on Greek REScoops from an SNM approach aims to explore the internal niche dynamics of ENC, examine the barriers that block a further systemic-change influence on current regimes, and recognize opportunities, networks, policies, institutional mechanisms and resources to enhance and shield a successful niche-building. However, we have to acknowledge that a sustainable strategic vision is not the rule for local energy projects, while some of them are satisfied with the fulfilment of local needs (Ziozas, 2020).

4.4 Case Studies

The research took place in three different case studies in the three Greek REScoops, namely the Sifnos Island Cooperative (SIC), the Electra Energy Cooperative (EEC) and the Energy Cooperative of Karditsa (ESEK). These very active EnCs represent three different types, which were selected as lighthouses; the first is insularly based focusing on a zero-energy island, the second is an intermediary actor with multiple activities nationally and the third one in the rural central Greek area has been fully operating for ten years (Ziozas, 2020).

Apart from the broad geographical distribution of the studied EnCs (Fig. 4.2), the specific cases provide a complete portrait of EnC activities and roles in the prism of niche analysis. In particular, the sample consisted of two citizen-led energy initiatives (i.e., ESEK and SIC),

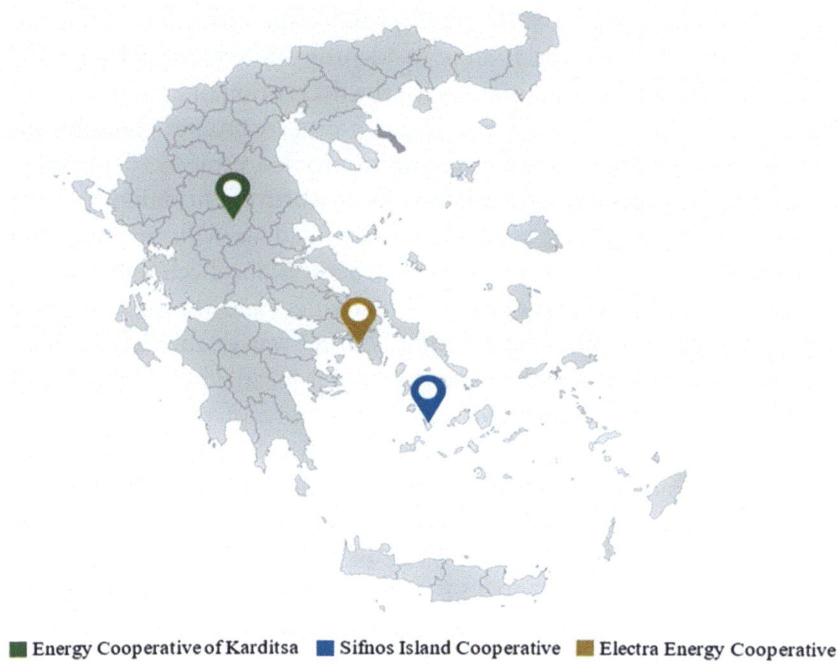

Fig. 4.2 Map of the examined REScoops in the Greek area

representing respectively the communities of Karditsa and Sifnos in this field, and one intermediary actor (i.e., EEC) dedicated to the development of local energy projects. A logical argument could be raised for the number of the case studies; however, according to Flyvbjerg, an extensive in-depth analysis of a single case can be more insightful from examinining the associated field as a whole (Flyvbjerg, 2006). Thus, it is important to present the profile of the studied energy cooperatives briefly in order to evaluate their contribution to the EnC niche holistically.

The choice of the examined cooperatives stemmed from the main research objectives to identify the niche phase of Greek EnCs, and to depict the influential capability of REScoop.eu Federation and EU Directives on it. The field research followed a qualitative approach with three in-depth semi-structured interviews from one representative of each REScoop (Table 4.1).

The examination of the recordings occurred with a narrative analysis central to the formation of social relations and experiences, and then, the analysed data fed into the SNM framework for the extraction of the particular results. This approach aims to identify underlying patterns and similarities among energy initiatives for the decryption of EnC as a niche. Our choice to develop qualitative research on case studies concerns the interactive and complex nature of community-led projects, demanding an in-depth engagement and analysis. It is worth mentioning that the specific sample of energy cooperatives provided us with an insightful diversity of legal forms, organizational structures, local economic status, integration of RES technologies, EnC activities, and the socio-cultural profile. *Energy Cooperative of Karditsa (ESEK)* was established in 2010 with the legal form of a civil cooperative and is located in the municipality of Karditsa. The cooperative was an initiative of the Development Agency of Karditsa SA (ANKA SA) and the Chamber of Karditsa, which met the unanimous support of public authorities and local stakeholders. Up to the present day, ESEK counts more than 476 members with main entry requirements, a cooperative share of 1000€ and proof of locality. The fundamental objectives of ESEK are the diffusion of RES technologies in the local energy mix aligned with the sustainable development of the local community. Initially, the general assembly of the cooperative had intended the design of an electricity production unit based on the

Table 4.1 Descriptive information of the three examined Greek REScoops

Greek REScoops	Community energy activity	Legal form	Members	Renewable energy source	Started	Interviewer	Citation code
Sifnos Island Cooperative (SIC)	Local energy initiatives	Civil Cooperative	>100	Wind, Hydro, Solar	2013	Member	[CEI]
Electra Energy Cooperative	Intermediary (Advocacy, Consulting)	Social Cooperative Enterprise	9	Solar, Wind	2016	Member (Vice-president)	[CE2]
Energy Cooperative ofKarditsa (ESEK)	Local energy initiatives	Energy Community (Former Civil Cooperative)	>476	Biomass	2010	ESEK Pellet Factory Manager	[CE3]

sustainable use of biomass and solar PVs. With the current inability to mobilize the necessary financing resources, the cooperative decided to construct a pellet production unit (Ziozas, 2020). The pellet production unit was built in 2017 and is located in an ESEK's property land of 22 acres, processing and standardizing local forest and agriculture biomass to produce solid biofuel of wood pellets with a capacity of 1100 tonnes of wood pellets per year. This project has mobilized a total supply chain of biomass in the local community from the harvest (i.e., forest cooperatives) until the production of the final product, activating environmental awareness as well as civil society engagement. The fruitful legislative framework on EnCs (Law 4513/2018) led ESEK to its legal transformation to an EnC. From an organizational governance perspective, the board of directors gathers every week to secure the viable operation of the pellet production unit and discuss the challenges of the EC. In addition, the General Assembly of the cooperative members is called once a year.

According to Geels and Deuten, the role of intermediaries is considered crucial for the successful nurturing of niches, while they constitute the missing link among isolated local initiatives and their guide for the further scaling-up to a global niche (Geels & Deuten, 2006). Warbroek et al. (2018) suggest "intermediaries" forms as a solution to build the capacities of local low-carbon energy initiatives. In the infant field of the Greek EnC niche, *Electra Energy Cooperative (EEC)* constitutes one of the few intermediary actors that contribute to the emergence of local sustainable energy initiatives. EEC is a social cooperative enterprise that was established in 2016 after the initiative of seven individuals, who all were sharing the same ideas for the development of EnCs and sustainable energy solutions in Greece. EEC envisions a clean, decentralized energy future, in which citizens will be the leading force of this energy transition due to their active participation. Based on this vision, its members have focused their efforts on the development and establishment of EnCs in the Greek context, providing services of advocacy, consultancy and policy support to a broad spectrum of involved civil society groups (i.e., municipalities, individuals, cooperatives, private enterprises). Among its activities, EEC, in collaboration with the Heinrich Böll Stiftung Foundation Greece, the Technical University of Crete, the Centre for Renewable Energy Sources and Saving (CRES), and the Winds of Renewal ("Anemos Ananeosis") cooperative, published the first guidebook exclusively dedicated to the building of EnCs in Greece (Kitsikopoulos et al., 2019). Moreover, it is designing the first EnC that will be based on the virtual net metering scheme through the use of solar PVs, namely "Hyperion", while already EEC has supported three EnCs to complete all the bureaucratic and technical procedures for their operation successfully. EEC became a member of REScoop.eu Federation in 2016, gaining fruitful experiences and knowledge from REScoops across Europe for its evolution.

Third, *Sifnos Island Cooperative (SIC)* was founded in 2013 on the Cycladic island of Sifnos under the legal form of a civil cooperative. The idea for its creation was inspired by its former Chair of the municipality council of Sifnos following a workshop in Brussels related to the development of local communities in the European context. Although its vision attracted a limited number of islanders at first, an advanced contact and later on fruitful cooperation with the Belgian energy cooperative *Ecopower* shielded Sifnians' expectations and led to the establishment of SIC. The cooperation aims to a green sustainable development of the island by exploiting the available local energy resources aligned with its simultaneously social and economic progress. The first effort for the RES penetration in the Sifnos energy mix occurred in 1989 after the study by the Public Power Corporation for a wind park of 1.26 MW capacity. A misunderstanding led to the suspension of the project until April of 2019. Sifnos, as with the majority of Greek islands, is not interconnected to the mainland electricity grid, covering its energy needs from a diesel power plant of 8.7 MW installed capacity. The economic and environmental unsustainability of this energy status quo led SIC to pose the issue of energy autonomy at the top of its agenda. Thus, after an open consultation for proposals, a hybrid power plant project consisted of a wind park of 11.5 MW and a pumped hydro storage plant with a hydropower capacity of 8 MW being chosen for the 100 per cent energy autonomy of the island. The feasibility study was published in 2016, and the project is currently was under revision by the Greek Regulatory Authority for Energy till today (mid-2021). The project is supported financially by three other REScoops with a total investment cost of 37 million euros. Moreover, SIC has also proceeded to a number of environmental activities, with the most successful one being that of the "Do it like a Sifnian" campaign. It was an initiative for the reduction of plastic bags in the daily activities of citizens, which managed to mobilize the civil society of Sifnos around environmental issues and to diffuse the multipurpose cloth bags into Sifnians' lives.

4.5 The Development of Internal Niche Processes Through the Lens of Greek REScoops

The status quo of the EnC in the Greek context is examined according to the SNM framework to highlight the dynamics and the evolution of this infant niche-building. Specifically, the evidence for the emergence of EnC niches looked from both REScoops' projects contribution ("Upward" flows) and associated intermediaries engagement ("Downward" flows) to the successful development of internal niche processes.

The expression *of expectations and visions* is considered the most critical internal process for successful niche-building, while it constitutes a guidance tool for the learning processes and allows the mobilization of valuable resources. It is normally expected there will be a dominance of downwards flows by the contribution of intermediary actors, while the local projects have less impact on this process. Even though the three REScoops have different approaches and rationales in their operation, they all share a clear vision for the building of the EnC concept in Greece towards a wider sustainable future for the involved societies.

An important finding is that the previous experience from members in the energy field as well as the evidence of an active community, played a crucial role in the robustness and influence of the respective expectations. For example, the experience of ANKA in community projects and the trustful cooperation with the Cooperative Bank of Karditsa facilitated the articulation of expectations among citizens and provided a higher quality to their vision (Ziozas, 2020). Another evidence is the registration of 476 members during its foundation day, consisting of a diverse mix of civil society groups (i.e., Chamber of Karditsa, local authorities). Similarly, EEC's members were all involved with EnC projects before the establishment of the cooperative. This fact made their expectations for EnCs more coherent, diffusing the same mature aspirations to the involved energy initiatives (Ziozas, 2020).

However, its intermediary nature does not leave space for a specific approach of EnC in terms of strict ideological lines, while EEC's objective is to be adjusted and understand the needs and difficulties of every local energy project that demands support. On the contrary, the case of Sifnos cooperative was based on the vision of a few islanders for a shift towards sustainability. The lack of knowledge and experience in the field of RES projects complicated the creation of a robust and specific vision of how SIC could approach their clean energy transition, though an exemplary case of an energy intermediary activity, that of REScoop.eu Federation, showed the pathway for Sifnians. Specifically, at the beginning *Ecopower* REScoop and later on, three more REScoops provided SIC with all the necessary resources, knowledge and policy support for the realization of their project due to long-term cooperation. This intermediary activity covered the lack of members' experience and shielded the vulnerable expectations for the energy project, diffusing a sense of realism and confidence. Despite the main downward flows of influence, SIC through some environmental activities (e.g., volunteer beach cleaning, the "Do it like a Sifnian" campaign) has pushed "upwards" the inspiration of involved citizens, strengthening more their expectations for an energy-autonomous Sifnos. Likewise, the construction and operation of ESEK's pellet production unit limited the uncertainty of Karditsa citizens, boosting their ambitions for further energy initiatives. Another insight from both ESEK and SIC development is an additional proposed aspect for the successful development of internal niche processes of building expectations that we called the "absorptive capacity" of visions. In particular, we observed in both cases that when members and the wider community "absorbed" the vision of EnC projects as their own and diffused its sustainability notion into their way of life, the rate of transforming expectations into projects was raised and the vision of a simple energy project could be transformed into a long-term strategy.

Specifically, the SIC members, despite the continuous challenges that the hybrid energy project has faced over the years, are loyal to and believers in its realization. Equally, the supply chain set up by ESEK in order to produce wood pellets spread the beneficial outcomes of such a collective initiative, diffusing the notion of solidarity and energy sustainability into citizens' behaviour. Finally, the evidence of SNM scholars that expectations have a dynamic nature was observed in all of our three cases (Mourik & Raven, 2006; Raven, 2012). ESEK changed its expectations for an electricity production unit from biomass to a pellet production unit, and now, in turn, they are designing the installation of a solar park (Ziozas, 2020). Similarly, SIC and the EEC changed and adjusted their coordination according to the current needs and challenges (i.e., bureaucratic, legislative, technological). However, this adaptability reinforced members' participation in the energy projects and facilitated the dissemination of information around the visions of the respective cooperatives.

Although the role of *shaping new social networks* in the development process of a robust niche is not very clear from the literature, their contribution is considered vital for the sharing of expectations and experiences, the mobilization of resources and the dissemination of knowledge (Seyfang et al., 2014). The new approach of EnC in Greece did not leave enough space for associated networking activities to flourish; however, the limited existence of a diverse set of intermediaries (i.e., universities, NGOs, institutes, municipalities, cooperatives) provide us with a remarkable case of a productive network in the EnC field (Warbroek et al., 2018). This evidence is reflected upon their contribution in collaboration with the Hellenic Ministry of Environment and Energy, included all the examined REScoops, in the creation of Law (4513/2018) on EnCs (Ziozas, 2020). A common finding for all three REScoops, which is also validated from the above-mentioned cooperation, is the dominance of a pre-existing network of intermediaries-to-projects linkages. Obviously, the internal networking characteristics vary for each case. For instance, ESEK has created a robust local network with public authorities, private investors, local institutions and citizens in the area of Karditsa that has been maintained from its foundation until the present. Mainly, this network includes informal relations, seminars and workshops among the involved local groups, aiming to secure the necessary social and financial resources for the project development. Moreover, the local-based network is vital for ESEK, while its final product of wood pellets is mainly addressed to the market and citizens of Karditsa. Thus, even though the

REScoop of Karditsa has developed a broad and deep network of social contacts, its influence is limited to a local level. Similarly, SIC networking activity is based on local contacts and friendships from the island of Sifnos, emphasizing the new generation of islanders through continuous educational actions (Ziozas, 2020).

However, in contrast to the ESEK case, the lack of local deep networking links in terms of mobilizing resources and the problematic SIC–state contacts (i.e., bureaucratic and legislative barriers) led SIC to a fruitful network-building with REScoop.eu federation. Creating personal relations with other REScoops members raised organizational and financial capital for Sifnos co-op, providing a wider influence of the Sifnos project to the EnC niche at both the European and national level. Specifically, the REScoop.eu Federation promoted an interactive video of Sifnos initiative as a worldwide paradigm for RES diffusion in insular areas. Also, the SIC hosted an international conference in Sifnos on "Energy Cooperatives, how they can be promoted in the island regions of Greece" in 2014, and one year later presented the plan for the energy autonomy of the island in the European Parliament. Nevertheless, the European networking activity of Sifnos co-op cannot overshadow the local social relations among citizens of Sifnos, while as a member of SIC commented, "the social acceptance of RES technologies and of our initiative as a whole from islanders is the basis of our effort". On the other hand, EEC as an intermediary actor has developed a broad network with projects and other intermediaries from the EnC field in both Greece and Europe. Indicatively, EEC has contributed to over thirty events (i.e., workshops, seminars) for the promotion of the EnC concept in Greece since 2018. However, the cooperative model of this effort, in combination with the weak civil society engagement to EnC initiatives, restrict the flow of human and financial capital dramatically to EEC. Overall, another fruitful finding from the examined cases is the missing networking link between the building of local social contacts and the establishment of formal relations at a European level. This link concerns the national layer of a dedicated network to the EnC niche that is characterized as vital from all three interviews. The creation of a national federation of energy cooperatives or a relevant networking platform of energy co-ops is considered a crucial step for the survival of the EnC effort in Greece (Ziozas, 2020).

Finally, the progress of *learning processes* is important for the critical first-niche stages of experimentation and for the formal identification of the respective niche (Mourik & Raven, 2006). According to SNM literature, the nature and depth of social networks determine the development of learning processes (Geels & Deuten, 2006); a statement that was validated from our research results. All our cases confirmed the immature and infant field of EnC in Greece from the dominance of first-order learning from both "upwards" and "downwards" flows of influence. Especially for both SIC and ESEK cooperatives that run energy projects in progress, the accumulation of data and findings is fundamental for their development phase. ESEK, building a broad social network in the community of Karditsa, had the opportunity to create a database and reports (i.e., amounts and quality of agricultural and forest waste, the environmental impact of the processes, supply chain of forest cooperatives) from the areas that were necessary for the successful operation of the pellet production unit. The circulation of experiences and "know-how" practices that emerged from the project were disseminated mainly through local public meetings, the general assembly, educational seminars, workshops, articles and technical reports. Similarly, SIC used the same techniques for the transfer of knowledge and expectations from its activities; however, the robust network with the REScoop.eu Federation and other European cooperatives boosted their influence and status (e.g., participation in Clean Energy Islands reports, published papers). Overall, for both energy projects, we observed an increased density in the production of first-order learning at the local level of their activity, although their presence and influence on a national level remains weak (Clean energy for EU islands).

On the contrary, the EEC, without the commitment to a single energy project, had the space to produce more generic lessons and findings for the EnC field. Among its reports and sharing of experiences from local initiatives, EEC contributed to the publication of the first handbook for the step-by-step building of an EnC in Greece. Moreover, the case of EEC confirmed to us that the broader the networking activity of an EnC intermediary is, the higher is the number of relevant second-order learning cases. The collaboration of EEC with Greenpeace to the production of an intermediating tool between EnCs and proposed investors, as well as its developing proposal for a national federation of EnCs in Greece, are evidence of the creation of new adjusted knowledge to the future needs of EnCs. It is also worth noticing that all of our cases were involved gradually to the aggregation of knowledge and practices from wider environmental activities irrelevant to the EnC niche. Although these learning outcomes are not connected directly to ENC, they set the foundations for a deep understanding of the sustainability notion.

4.6 Discussion and Conclusions

Barriers exist in the current regulation scheme, since it is completely impossible to form small schemes of collective energy self-production by households practicaly. This is due to the operating costs of the mandatory start of work with the tax offices. As a result, the target of 600 MW new self-generating schemes by 2030 of the National Plan for Energy and Climate is not feasible.

Even though the current EU policies recognize the particular characteristics of EnCs, such as their social, environmental and economic advantages, the current Greek legislative and regulatory framework is not capable of assisting their fruitful progress. The current framework is complicated and allocated in many laws and ministerial decisions, which is a barrier to the creation and operation of RES projects by EnCs.

Furthermore, costly and complex administrative procedures continue to apply, while also EnCs face several problems regarding access to finance and information.

Additionally, EnC is used as a tool by private investors, who exploited the relevant opportunities of real initiatives of the local communities. As a result, even the requirement of proximity of EnCs, which aimed to strengthen local communities, has become a disadvantage compared to private actors, who do not have to overcome geographical constraints.

So, according to Law 4759/2020 all EnCs should compete with private investors in the energy market bids. However, for the time being, the adoption of completely competitive processes for all market players will inhibit the operation of EnCs, and, in some cases, will result in their withdrawal, as already noticed in the other EU Member States like Germany and France.

The authors have discussed these issues with several of the involved organizations. This led to formulating the following future recommendations (to policy-makers):

- There is a need for deep assessment of barriers and development opportunities of EnCs (potential, barriers, financing schemes).
- It is suggested that proper EnC financing schemes are required and should hence be established; i.e., bank loan guarantees to finance virtual net metering, seed-funding to cover creation expenses, and sources from a recovery fund, such as in the Netherlands.
- It is suggested that there is a prioritization in large islands (e.g., Crete, Rhodes, Lesvos) to cover up to 50 per cent of the peak demand; and for smaller off-grid islands with peak power less than 10 MW up to 100 per cent.
- Legislative modifications are required to allow EnCs with less than sixty members (or with the participation of local authorities) to participate in the competitive RES project bid processes after 2021.
- There is a need to design an operational model of collective self-production and virtual net metering, which is appropriate for small groups.
- It is suggested to develops plans for the future participation of EnCs in offshore RES projects.

Although EnCs bring additional benefits to the promotion of a green energy transition compared with those of private investors, the Greek EnCs currently face crucial problems. They need sufficient support so that they can participate in the energy market as an equivalent market player without discrimination, according to EU Directives.

Acknowledgements The authors of this chapter would like to particularly thank the community energy practitioners of each Greek REScoop for their enthusiasm and time in taking part in this study. Specifically, we would like to thank separately each cooperative for their contribution per se, namely the

Sifnos Island Cooperative, the Electra Energy Cooperative and the Energy Cooperative of Karditsa. This research approach was partially supported by the H2020—REScoop Plus project (grant agreement No. 696084). The sole responsibility for the content of this publication lies with the authors.

References

Businessregistry.gr. (2020). *Public Data*. [online] Available at: https://www.businessregistry.gr/publicity/index. Accessed 1 Mar 2021.

Clean Energy for EU Islands https://euislands.eu/.

Flyvbjerg, B. (2006). Five Misunderstandings About Case-Study Research. *Qualitative Inquiry, 12*, 219–245.

Gancheva, M., O'Brien, S., & Crook, N. (2018). *Models of Local Energy Ownership and the Role of Local Energy Communities in Energy Transition in Europe.* Commission for the Environment, Climate Change and Energy. European Committee of the Regions.

Geels, F., & Deuten, J. (2006). Local and Global Dynamics in Technological Development: A Socio-Cognitive Perspective on Knowledge Flows and Lessons from Reinforced Concrete. *Science and Public Policy, 33*(4), 265–275.

Geels, F. W. (2002). Technological Transitions as Evolutionary Reconfiguration Processes: A Multi-Level Perspective and a Case-Study. *Research Policy, 31*(8–9), 1257–1274.

Geels, F. W., & Raven, R. (2006). Non-Linearity and Expectations in Niche Development Trajectories: Ups and Downs in Dutch Biogas Development (1973–2003). *Technology Analysis & Strategic Management, 18*(3–4), 375–392.

Greek Law 4513/2018. (2018). *Energy Communities and Other Provisions*. 23 January 2018.

Kitsikopoulos, D., Damasiotis, M., Korma, E., Metaxa, K., Barkouta, D., Savvakis, N., & Tsoutsos, T. (2019). *Building Energy Communities*. Energy in the hands of citizens, Heinrich Boll Stiftung.

Law 3852/2010. (2010). *New Architecture of Self-Governance and Decentralized Administration*.

Law 4555/2018 (Government Gazette 133 A'/2018). *Reforming the Institutional Framework of Local Administration*.

Law 4759/2020. (2020). *On the Deficit of the RES Special Account (ELAPE)*.

Mourik, R., & Raven, R. (2006, December). *A Practioner's View on Strategic Niche Management: Towards a Future Research Outline*. Energy Research Centre of the Netherlands.

Oecd.org. (2019). *Subnational Governments Around the World – OECD*. [online] Available at: https://www.oecd.org/regional/regional-policy/sngs-around-the-world.htm. https://www.oecd.org/regional/regional-policy/profile-Greece.pdf. Accessed 28 Feb 2020.

Raven, R. (2012). Analysing Emerging Sustainable Energy Niches in Europe: A Strategic Niche Management Perspective. In G. Verbong & D. Loorbach (Eds.), *Governing the Energy Transition: Reality, Illusion or Necessity?* (pp. 125–151). Routledge.

RESCOOP plus. http://www.rescoop-ee.eu/rescoop-plus. Accessed 28 Feb 2020.

REScoop.eu, Greenpeace, WWF Hellas and Electra Energy. (2021). *Development of Energy Communities in Greece: Challenges and Recommendations*. https://www.rescoop.eu/news-and-events/news/executive-summary-development-of-community-energy-in-greece-under-pressure. Accessed 1 Mar 2021.

Schot, J., & Geels, F. (2008). Strategic Niche Management and Sustainable Innovation Journeys: Theory, Findings, Research Agenda, and Policy. *Strategic Management, 20*(5), 537–554.

Seyfang, G., Hielscher, S., Hargreaves, T., Martiskainen, M., & Smith, A. (2014). A Grassroots Sustainable Energy Niche? Reflections on Community Energy in the UK. *Environmental Innovation and Societal Transitions, 13*, 21–44.

Sifakis, N., Daras, T., & Tsoutsos, T. (2020). How Much Energy Efficient are Renewable Energy Sources Cooperatives' Initiatives? *Energies, 13*, 1136.

Sifakis, N., Savvakis, N., Daras, T., & Tsoutsos, T. (2019). Analysis of the Energy Consumption Behavior of European RES Cooperatives Members. *Energies, 12*, 970.

Smith, A., Hargreaves, T., Hielscher, S., Martiskainen, M., & Seyfang, G. (2015). Making the Most of Community Energies: Three Perspectives on Grassroots Innovation. *Environment and Planning A: Economy and Space, 48*(2), 407–432.

Smith, A., & Raven, R. (2012). What Is Protective Space? Reconsidering Niches in Transitions to Sustainability. *Research Policy, 41*(6), 1025–1036.

Tampakis, S., Arabatzis, G., Tsantopoulos, G., & Ioannis Rerras, I. (2017). Citizens' Views on Electricity Use, Savings and Production from Renewable Energy Sources: A Case Study from a Greek Island. *Renewable and Sustainable Energy Reviews, 79*, 39–49.

Various. (2002). RE Islands: RE Hybrid Systems Integration. *Refocus, 3*(4), 54–57.

Verbong, G., & Loorbach, D. (2012). Introduction. In G. Verbong & D. Loorbach (Eds.), *Governing the Energy Transition: Reality, Illusion or Necessity?* (pp. 1–23). Routledge.

Warbroek, B., Hoppe, T., Coenen, F., & Bressers, H. (2018). The Role of Intermediaries in Supporting Local Low-Carbon Energy Initiatives. *Sustainability, 10*(7), 2450.

Ziozas, N. (2019). *The Role of Civil Society in Community Energy niche: A Case Study of Greece.* Master Dissertations, Science Research Policy Unity (SPRU), University of Sussex.

Ziozas, N. (2020). Notes and Reports from in situ mission to the Energy Communities.

5

The Community Energy Sector in Italy: Historical Perspective and Recent Evolution

Chiara Candelise and Gianluca Ruggieri

5.1 Introduction

The Italian community energy sector was one of the main actors in the development of the electrical system in the early 1900s. After the nationalization of the system, a quasi vertical integrated monopoly was set up: the community approach was then limited to very specific local situations until the early 2000s. The liberalization of the energy markets as well as

C. Candelise (✉)
GREEN, the Centre for Research in Geography, Resources, Environment, Energy and Networks, Bocconi University, Milan, Italy

Imperial College Centre for Energy Policy and Technology (ICEPT), Imperial College London, London, UK
e-mail: chiara.candelise@unibocconi.it

G. Ruggieri
Dipartimento di Scienze Teoriche e Applicate, Università degli Studi dell'Insubria, Varese, Italy
e-mail: gianluca.ruggieri@uninsubria.it

© The Author(s), under exclusive license to Springer Nature Switzerland AG 2021

F. H. J. M. Coenen, T. Hoppe (eds.), *Renewable Energy Communities and the Low Carbon Energy Transition in Europe*, https://doi.org/10.1007/978-3-030-84440-0_5

the introduction of the feed-in tariff made possible the development of some new experiences earlier than in other European countries, although limited in number and dimension.

The main question in this chapter is as follows: How did Renewable Energy Communities in Italy evolve, and what we can learn from these experiences for the realization of the Clean Energy Package in Italy? A new spring for the Italian community energy sector may be announced by the implementation of the EU Directive introducing Renewable Energy Communities and Citizen Energy Communities. A number of pilot projects are currently under development. Whether this will lead to a revival of local small-scale experiences or will reinforce larger Italian renewable energy initiatives or bring new actors on the scene is an open question worth of analysis and discussion.

5.2 Brief Historical Introduction

The first energy communities in Italy date back to the end of the nineteenth century when local cooperatives were founded to support the electrification of settlements in rural areas due to a lack of national grids and an overall prioritization of urban areas. The electrification in Italy started in 1883 in Milan where the coal thermoelectric plant of Santa Redegonda was built and in the following years a number of private initiatives were developed in urban contexts side by side with municipal utilities, promoted by a national law in 1903 (Cringoli, 2017). In rural, alpine regions local cooperatives were established to provide electricity to their members. They were developed around the construction of hydroelectric plants and the development of a local distribution grid (e.g., Società per l'Illuminazione Elettrica in Chiavenna, 1894) (Scaramellini, 1995). In 1962 ENEL, the Italian national utility, was founded and several of these energy cooperatives were included in this national quasi vertical integrated monopolistic utility during the nationalization of the electricity system. Some of the energy cooperatives refused to join and continued their own activities even after the liberalization process started in 1999 with the implementation of the EU Directive 96/92/CE concerning common rules for the internal market in electricity (Di Gaspare, 2003). These so-called "historical" energy cooperatives are still operating and have been expanding their activities by deploying additional renewable energy capacity from other generation technologies such as photovoltaics or biomass. There are reported to be around seventy-three of these, accounting for 80,000 clients, serving sixty municipalities and producing 500GWh of renewable electricity (Confcooperative, 2021). However, these are not initiatives that can be easily replicated as they have been allocated a special legal status that allows them to own and manage local distribution grids (Spinicci, 2011).

Since 1999 the liberalization process has paved the way for a deep restructuring of the energy sectors (both gas and electricity). Markets have been established in substitution of the national monopoly. A number of different actors have entered the energy markets, both in the production and in the supply sectors. End consumers were given the possibility to choose between different suppliers, including the energy intensive industry. Since 2007 all end consumers are able to switch their supply contracts if they want to.

In addition to liberalization, the transition towards decentralized energy systems led by renewable energy production and policy support have left space for smaller-scale investments in energy assets, opening up the possibility to new actors such as municipalities, SMEs and citizens to enter the energy market. The energy users are allowed to turn into "prosumers" or co-providers of energy services. In Italy, this has been particularly evident since 2001, when national support mechanisms have been introduced to support the progressive deployment of the photovoltaics (PV) sector. First, the 10,000 PV roofs program enabled 4.5 MWp of PV installation as privately owned rooftop projects across the country. Later, a feed-in tariff scheme was introduced in 2005 and was then renewed in five subsequent periods, leading to a considerable expansion of the PV installation capacity in Italy in the following years (Gangale & Caminiti, 2005). More than 500,000 PV plants (for a total of more than 18 GWp) were eventually installed between 2005 and 2013, most of them between 2009 and 2013 (GSE, 2016).

In this period of deep evolution, new opportunities emerged for citizen collective initiatives in the form of "community energy" (CE). A new wave of energy community initiatives in Italy began around 2008. They mostly concern initiatives focused on the development of renewable energy production facilities and, most of all, differentiate themselves from Italian historical cooperatives as they do not benefit from their special legal status and cannot own local distribution networks.

5.3 New Wave of Community Energy in Italy: the 2008–2013 Period

Italy was among the first European countries to develop CE experiences in the 2000s, and between 2008 and 2013 several energy community initiatives were established that aimed at engaging people in the energy transition. A systematic review (Candelise & Ruggieri, 2020) of these new paradigms of citizens' financial and ownership involvement in energy initiatives was developed to explore their characteristics, dynamics of development and the forms and level of citizens' involvement. The review identified seventeen energy community initiatives, a relatively small, but nonetheless rather complete sample. It provides a snapshot of the Italian CE sector prior to the very recent and ongoing sector evolution due to initial implementation of EU Directive (see Chap. 2 and Sect. 5.4) and it highlights some trends in their characteristics and in the conditions for their development. The review includes initiatives:

1. that imply a form of citizens ownership or financing of an energy project, as well as control over the initiatives.
2. where citizens directly benefit from the outcomes of the initiative.

At least in principle these characteristics are in line with the definitions of Renewable Energy Communities and Citizen Energy Communities provided by the European directives (see Chap. 2 and Sect. 5.4) although direct electricity supply was not possible at that time.

5.3.1 Dynamics of Creation and Organizational Structures

The development of CE initiatives may be started by citizens or other types of grassroots green organizations or in contrast by a public institution or an established market actor. The Italian evidence shows a majority of the initiatives proposed through a top-down approach (i.e., where a third-party institution leads the development of the project and facilitates citizens' involvement); of those, five have been proposed by a municipality and seven by a private company or a municipal utility. Only five initiatives have been initiated with a bottom-up approach (i.e., characterized by strong involvement of citizens or other types of grassroots organizations in the initiation and development of the project).

Even when they do not initiate the projects, local authorities may have an important role also as facilitators in several cases, by providing the assets to develop the initiative, such as public-building rooftops, or by creating the local regulatory and financing framework conditions to allow it. This reinforces literature views on their potential key position in facilitating energy transitions and influencing local energy system change (Hannon & Bolton, 2015; Rutherford & Jaglin, 2015; Legambiente, 2016).

As also experienced in other countries (Creamer et al., 2019; Yildiz, 2014) the legal structure adopted may include limited companies, non-profit associations and cooperatives, which account for about 60 percent of the sample. Cooperatives are generally deemed to provide the best institutional framework for locally owned and participatory approaches to renewable energy projects. They encompass both the social and economic dimension in their scope and are characterized by a 'one head one vote' decision-making process, with the aim to provide higher levels of co-determination. However, evidence from Italian samples show that the level of citizens' participation and control over decision processes is not generally determined by the legal form adopted and the relative internal governance as defined by national laws and regulations. For example, there are initiatives structured as limited companies which have been developed with higher involvement of the citizens than some cooperatives in the sample.

Independently by the legal status adopted, initiatives proposed by public and private companies and with a strong top-down approach have been developed with lower involvement of citizens, and their organizational structure implies lower citizens' co-determination. This also emerges from the evolution of the financing structure adopted: four projects have been initially financed through some form of project financing (three cooperatives proposed by a company and the project proposed by a municipal utility) and only in a second phase they were opened to citizens' financing. Instead, initiatives promoted by communities and municipalities have been founded and developed through the direct financial contribution of citizens.

5.3.2 Type of Activity and Timing

The great majority of CE initiatives included in the review have electricity generation as main activity and fifteen out of seventeen initiatives have developed PV systems as their primary activity (the other two are an electricity supplier and one wind farm). Indeed, one of the main factors of the development of CE projects in Italy has been the introduction of the generous and uncapped feed-in tariffs (FiT) for photovoltaic (PV) systems. Between 2008 and 2013 the FiT scheme has been guaranteeing fixed long-term tariffs and net-metering to PV system owners. Such strong policy support, combined with remarkable reductions in PV modules and installation costs since 2010, have made PV investments quite profitable and relatively low risk in the wider context of the Italian energy sector.

This window of opportunity was particularly important for proponents generally not equipped to deal with large, complex and high-risk project development in the energy sector. Moreover, they provided enough resources to benefit both the citizen investing in the projects and the partner providing the rooftop for the installation (in most cases schools or other public buildings). While the incentives were used to reward the investors, the building owner could use the energy in a self-consumption scheme, decreasing its energy bills.

Nonetheless, the Italian CE sector has been mainly characterized by the development of rather small, 'ad hoc' initiatives with a strong local focus. While PV systems installed vary in size and application, the majority are small/medium-size projects, more easily developed and financed by actors with a low degree of experience in the energy sector.

The national cap of 6.7 billion € set for the FiT scheme was reached in 2013, leading to the discontinuity of the FiT support scheme. As a consequence, the Italian PV market has contracted moving from 3.5 GW/year of installed PV between 2008 and 2013 to 385 MW/year in the period between 2013 and 2018 (GSE, 2020). Most of the CE renewable energy plants have in fact been developed between 2008 and 2013. Afterwards, only a few CE initiatives have been developing new energy projects, widening their scope to a national scale. Moreover, those still operating after 2013 have rarely developed new renewable energy plants and mostly focused their activity on the secondary market, purchasing operating PV plants which are still benefiting from the FiT support. We will further analyze these initiatives in the following section.

5.3.3 Economic and Social Impacts of CE Initiatives

All the CE initiatives surveyed involve a form of financing or ownership from members against which a monetary return is offered. The returns on investment offered to citizens can vary quite substantially, from 1 percent to 8 percent. Such a variation may seem particularly striking considering that most initiatives have been investing in the same energy technology—PV systems. This can be partly explained by the size and typology of the PV plant: larger ground-mounted plants allow higher economies of scale in the investment (both in terms of initial capital costs and transaction costs) and therefore higher returns than smaller roof-mounted systems. However, what makes a stronger impact on the monetary returns offered to citizens is the typology of the initiative. Under this perspective, two distinctive typologies of initiatives may be identified.

A first group includes initiatives whose primary activity is the production of electricity from a renewable energy plant (in most cases PV). Their main objective is the distribution among members of the revenues accruing from the operation of a renewable generation project. The revenues are generally distributed as a direct payout or, in some cases, through electricity bill savings or a combination of both. Since it was not possible to provide to the participant the electricity produced in the plants, an electric supply company is involved in the scheme, which may acquire and then sell it to the investors. This complex virtual self-consumption scheme can be useful in distributing economic benefits as a bill discount (avoiding any taxation on profits). These kinds of initiatives have generally developed a single renewable generation project as a unique primary activity. Higher financial returns, on average around 6–8 percent, are offered by these initiatives.

A second group includes initiatives which are set to offer other energy and social services to benefit both cooperative members and wider local communities. Their activities go well beyond developing renewable energy plants and aggregating citizens around the relative financing and ownership. Lower financial returns on the investment are offered in this case, as they tend to have more complex organizational structures. Revenues from investments in renewable generation projects can be distributed across a wider set of activities including those that do not generate monetary benefits for their members, including domestic energy efficiency audits and consultancy, collective purchasing of energy services, as well as wider community development schemes such as information campaigns or activities with schools.

5.4 Evolution of the Sector in a Context of Low Policy Support: È Nostra, WeForGreen and Energia Positiva

Since the progressive discontinuity of the FiT support in 2013 and the reintroduction of market-based support, the scaling-up of the EC sector either by developing large plants or replicating smaller projects has proven in most cases to be challenging. The small-scale model became no longer profitable and sustainable, and new approaches were needed. Only three initiatives managed to overcome this critical phase and to enlarge their activities: Retenergie/È nostra, WeForGreen and Energia Positiva (Candelise & Ruggieri, 2020).

Retenergie, founded in 2008, is among the first initiatives of the new wave of EC in Italy. È nostra, founded in 2016, was the first Italian EC to supply green electricity to its members. In 2018, Retenergie merged into È nostra, thus creating a cooperative able to provide both production and supply of renewable electricity and to offer energy service to a national community of prosumers. At the end of 2020 the new È nostra includes 7000 members/customers and provides 20 GWh/year. ForGreen is a limited company born as a spinoff of an Italian multi-utility in 2010 with the aim of developing PV systems and energy efficiency services. In 2015 ForGreen developed a new cooperative, WeForGreen Sharing. The cooperative now works as an umbrella for at least five different EC initiatives focused on the development of a single renewable energy plant, with the aim of supplying green electricity to its members through an electricity bill saving scheme, which represents a relevant component of the guaranteed return. Energia Positiva started its operations in 2016 and by the end of 2019 had developed fifteen projects, ten of which are PV plants benefiting of FiT support acquired on the secondary market (for a total of 1.5 MWp approximately). Energia Positiva reported a total investment almost 5 M€ by 415 members (average investment of about 12,000€ per member). To become members of Energia Positiva individuals invest in quotas of the cooperative, which are linked to specific projects in order to become owners of a "virtual renewable energy plant". The return on the investment is guaranteed with a direct discount on the electricity bill.

The three initiatives have some similar features. First, they all increased their activities by focusing on larger size projects and/or developing multiple projects. They enlarged the territorial scale of their activities, both by developing projects in different locations across the country and by involving members at a national scale. They thus managed to achieve economies of scale in their activities, which allowed them to involve and hire professionals as permanent staff and progressively enhance the services provided. They overcome the lack of incentives for new plants by focusing on the secondary market of PV, thus investing in plants with higher profitability and lower risk because they were still benefiting from the FiT support. More recently Energia Positiva and È nostra have started differentiating projects activities by also developing energy efficiency projects and a wind project.

As a result, although on the one hand they lost the local dimension in project development, on the other they could integrate the proposition offered to their members with a model that combines participative renewable energy production with provision of green electricity. They managed to do so in different ways. 'È nostra' is a proper electricity supplier. 'Energia Positiva' and 'WeForGreen' instead subscribed an agreement with other green electricity supply companies linking the rebate on electricity supply proportionally to the investment of their members in the cooperatives. Nonetheless, the three cooperatives have developed a sort of 'virtual' prosumer model, which allows citizens across the entire Italian territory to support and participate in their renewable energy production projects while also directly consuming green electricity.

A closer look at the three initiatives also highlights different approaches in their development and growth over time. Evidence presented on the Italian CE sector has highlighted two typologies of initiatives (see previous section).

Energia Positiva and WeForGreen belong to the first typology whose primary activity is the production of electricity from a renewable energy plant. Indeed, Energia Positiva's growth seems to be rooted in the successful replication of a model in which the investments in single renewable energy projects are shared among members through a sort of quota system in exchange for participation in the revenues accruing from them (despite being in the form of electricity bill savings). This modular approach has allowed a constant grow over time of the initiative, which has been steadily developing projects and has recently raised more finance through a crowdfunding campaign to support further expansion. A very similar approach has been followed by WeForGreen which has developed a smaller number of projects, but of larger size, probably thanks to the fact that they are supported by an energy company. These typologies of initiatives are typically able to offer higher financial returns and, among all those mapped within the Italian CE sector, Energia Positiva and WeForGreen offer the highest at around 8–9 percent.

È nostra (as well as Retenergie) was developed to offer other energy and social services to benefit both cooperative members and wider local communities, and belong to the other typology. Retenergie structure was more complex than Energia Positiva and WeForGreen, both in term of its activities (as it combined renewable electricity production projects with energy and community services) and in its financial structure and citizens' engagement process. Revenues generated by investments in renewable generation projects have been redistributed across a wider set of activities which did not generate direct monetary benefits for their members. This has resulted in lower returns offered (ranging from 0 percent to 3 percent). In addition, the merging with È nostra has created a unique entity which offers not only collectively owned renewable energy production and energy services, but also an electricity supply. È nostra is also involved in organizing training, information and education activities. Compared to Energia Positiva and WeForGreen, this more complex structure would make quick replication and upscaling of the model less viable.

In conclusion, it is clear that only initiatives able to move from the small and local dimension toward a national and higher scale managed to grow in the context of contracted policy support. The different scaling-up strategies are affected by the early choices on the initiatives' activities and organizational structures made since the founding stage. Energia Positiva and WeForGreen follow a growth path more focused on serving mutual interest (i.e., serving the interest of their members) while the Retenergie/È Nostra scale-up has been more informed by general/public interest (i.e., serving the broader interest of society).

5.5 Current Situation and Possible Ways Forward for Community Energy in Italy

Italian CE initiatives discussed in the previous sections have been developing between 2008 and 2019 without any direct policy support. As discussed, the sector has indirectly benefited from the Italian FiT tariff scheme, whose primary objective was the expansion of PV technology deployment, hence it was not explicitly designed to support EC. This reflects a general trend in Europe, where the recognition of ECs within the energy policy framework have been limited to some specific countries initiatives (such as Scotland or Denmark (Bauwens et al., 2016; Cobut, 2021)). This was until two EU directives provided a formal definition and a relative regulatory framework for EC. As discussed in Chap. 2, two definitions of EC have been introduced: Citizen Energy Communities (CECs) contained in Directive (EU) 2019/944 (recast Electricity Directive), and Renewable Energy Communities (RECs), contained in Directive (EU) 2018/2001 (the recast Renewable Energy Directive).

In Italy CEs have recently attracted the attention of the legislator, with a progressive recognition of their potential role within the Italian energy system. ECs were first mentioned within the Italian regulatory framework by the Italian Energy Strategy in 2017 and, subsequently, by the National Energy and Climate Plan in 2018. However, they were both legislative framework documents, which did not imply any concrete measure to support the implementation of CE initiatives in the country. In 2018, the Piedmont region implemented a law on ECs, which has mainly been a declaration of intent, although politically relevant, being the first legislative initiative explicitly dedicated to the Italian CE sector. In 2019, two other Italian regions, Apulia and Sardinia, have both implemented a similar regional law on ECs.

Since the end of 2019 the process of national implementation of the two EU directives has created momentum for the development of specific policy support to the CE sector in Italy. In February 2020, as an initial step toward the national implementation of the EU Directive, Law 8/2020[1] provided the first formal definition within the Italian regulatory framework for Renewable Energy Communities (RECs). RECs are defined as entities made by members which collectively self-consume electricity produced by local renewable generation. The law also defines some boundaries in the implementation of RECs, by providing that the renewable energy plant serving the REC can be up to 200 kWp and

[1] Law 28 February 2020, n. 8 "Conversione in legge, con modificazioni, del decreto-legge 30 dicembre 2019, n. 162, recante disposizioni urgenti in materia di proroga di termini legislativi, di organizzazione delle pubbliche amministrazioni, nonché di innovazione tecnologica."

giving a strict definition of "locality", i.e., requiring that RECs can only include members linked to the same low-voltage distribution sub-grid.[2]

Law 8/2020 has paved the way for the first direct policy support to EC in Italy. In August 2020, the Italian Energy Authority (ARERA, Autorità di Regolazione per Energia Reti e Ambiente) defined the incentive regime in support of RECs.[3]

The RECs are entitled to a direct incentive per kWh of electricity self-consumed (110 €/MWh) and the partial refund of the transport tariffs according to a cost-reflective approach (10 €/MWh). The total incentive would therefore amount to 120 €/MWh. The incentive is granted only for the kWh of electricity which are consumed simultaneously with electricity production. Hence, it is designed to incentivize consumer behavior that would maximize local consumption of renewable energy produced and thus minimize export to the grid. The electricity that is not self-consumed within the REC is sold to the grid at a zonal price.

The regulatory implementation for this policy support was set in December 2020, with the publication by GSE (Gestore dei Servizi Energetici, the Italian national energy services Operator) of the "Technical rules for access to the shared electricity incentive service",[4] which indicates practical rules and procedures to access the policy support. It was therefore in 2021 that these new developments in the Italian EC sector were beginning to unfold.

It is relevant to point out how the definition and the regulatory framework developed in Italy in support of REC clearly originates from the definitions of REC and CEC provided by the EU directives and their "dual purpose": an "energy purpose", i.e., the need to increase local self-consumption within the energy system, and the purpose of achieving

[2] The Law 8/2020 also defines Collective Self-Consumption (CSC) projects, whose main difference from RECs is that their member have to live in the same building, while RECs include members linked to the same low voltage distribution sub-grid.

[3] ARERA 2020 Delibera 04 August 2020 "Regolazione delle partite economiche relative all'energia elettrica condivisa da un gruppo di autoconsumatori di energia rinnovabile che agiscono collettivamente in edifici e condomini oppure condivisa in una comunità di energia rinnovabile" 318/2020/R/eel.

[4] GSE 2020 "Gruppi di autoconsumatori di energia rinnovabile che agiscono collettivamente e comunità di energia rinnovabile—Regole tecniche per l'accesso al servizio di valorizzazione e incentivazione dell'energia elettrica condivisa."

social, environmental and economic benefits for the local communities through citizens' participation. As mentioned above, the Italian policy support is indeed designed to support the "energy purpose", by incentivizing simultaneous self-consumption and hence reducing grid exports and the relative grid impacts and management needs. In addition, following on from EU directives, RECs are conceived to support environmental, social and economic purposes. They are defined as follows:

- constituted by group of citizens, SMEs and local authorities living in the same local areas;
- their main objective has to be providing environmental, economic and social benefits to its community of members or to the local context where the REC operates rather than financial profits;
- although no specific requirement is provided by the law, in this perspective they should take the form of third sector legal entities (e.g., associations od cooperatives).

Another relevant element of the Italian regulation is the fact that the developer as well as the owner of the renewable energy plant serving the community could be a different legal entity than the REC itself. This leaves space for the entrance in the EC market of third parties, such as ESCOs, and renewables developers as well as the existing ECs (presented in previous sections) as potential facilitators of the community and providers of technologies as well as of finance to support the development of RECs. As we will discuss later on, the definition of RECs leave space for several possible configurations, in particular depending on: who are the members composing the community (domestic, non-domestic, local authorities) and who is going to develop and own the plant serving the community.

The regulatory regime introduced by Law 8/2020 has been intended by the Italian legislator as an initial step toward a more complete and wider implementation of the EU directive. In the following we highlight a number of open issues that will be probably clarified in future years by the Italian regulator in the progressive full implementation of the EU directive provisions on both RECs and CECs.

5.5.1 Legal/Normative Open Issues and Expected Evolution

An open discussion is currently ongoing on the definition of "locality" (i.e., the limitation to the low voltage distribution sub-grid) and the maximum generation power for the renewable energy plant serving the REC (i.e., 200 kW). The overall aim of the regulation to incentivize local self-consumption to allow higher levels of penetration of intermittent renewable generation is not questioned. However, whether 200 kW is a reasonable maximum generation power or it is limiting the possible development of RECs is still an open question. The current debate seems to indicate a willingness of the regulator to lift this limit as well as to redefine "locality" in order to allow the implementation of RECs of larger geographical sizes. On the latter some of the options discussed are to use the municipality, the zip code, or a higher-level grid substation as new geographical boundaries of the RECs. Indeed, the intention of the regulator seems to move forward from the current provisional regulation to implement a wider and larger definition of REC. This would also go along with a definition of CECs in the Italian legislation, as Italy has not yet implemented the recast Electricity Directive.

5.5.2 Several Technical Aspects Can Influence the Actual Possibility of Creating Energy Communities

In the transition from a defined regulatory framework and incentive to the actual implementation of RECs a series of technical aspects have been emerging.

First, since the actual simultaneity of production and consumption must be verified, it is necessary that RECs' members own a smart meter able to record the consumption curve (i.e., 2G smart meters). However, the rollout of 2G smart meters in Italy is still ongoing (with a target of 85 percent of smart meters installed by 2024). This means that currently several potential REC members need to face the additional cost of smart meter installation in order to be able to join the REC.

Another technical concern is the lack of access to information on grid assets and maps for the wider public, as this information is available to the distributors only. Since only the final users connected to the same low voltage distribution sub-grid can be included in a REC, it would be important that the design of the grid should be publicly available without prejudice to the rights to privacy. This lack of access to data is at the moment slowing down the practical deployment of RECs. Furthermore, since distributors can also be promoters of RECs, the fact that they exclusively own the data and information on grid design could generate market distortions.

More generally, a wider concern is whether the current configuration of RECs and the relative regulatory framework would lead to a level of RECs deployment which would actually help in reducing grid congestion and easy higher levels of renewable generation penetration into the Italian electricity network. The first pilot projects and initial market developments should provide the initial evidence to support this argument, and potentially allow for a refinement of the regulation in order to achieve this objective.

5.6 Market Potential for RECs in Italy

The current regulation and policy support for REC is very new and the market has just begun to investigate implementation models for REC deployment. In general, the current regulatory framework leaves space for several potential configurations of RECs. They can vary in terms typology of end-users connected, which could be domestic or non-domestic customers as well as industrial districts. They can also differ in terms of dynamics of creation, financing and ownership, spanning between two extremes: a fully bottom-up approach and a top-down approach. In the former, the members organize and constitute the REC as well as develop and finance the renewable energy plant that serves it. In the latter, a third party is the promoter of the initiative and takes care of the development and financing of the renewable plant, which then serves the community of end-users organized under a no-profit entity.

A recent report by the Energy and Strategy Group (ESG) of Milan Polytechnic (ESG, 2020) has identified several possible configurations of RECs and derived three possible development scenarios for RECs over the 2021–2025 period:

- a low penetration scenario accounting for: 700 energy communities, 100 MW PV installed, 15,000 Households, 4000 offices and 200 SME;
- a medium penetration scenario accounting for: 26,000 energy communities, 3600 MW PV installed, 750,000 Households, 150,000 offices and 8000 SME;
- a high penetration scenario accounting for: 40,000 energy communities, 5400 MW PV installed, 1,200,000 Households, 200,000 offices and 10,000 SME.

Under ESG's hypothesis the difference between the lower scenario and the other two is the possible role of third parties, which would act as facilitators and providers of technologies and technical skills as well as of finance to support the development of RECs. Along with ESCOs, SME and small developers, these may include utilities as well as institutional investors (e.g., Cassa Depositi e Prestiti, SpA controlled by the Ministry of Economy and Finance that promotes the growth of the country and manages postal savings) or new market actors such as Poste Italiane (the Italian postal service provider) which recently announced its entry into the energy market and is present in many of the 8000 Italian municipalities.

The role of third-party energy market players as facilitators of RECs is therefore deemed to be an important factor to guarantee higher levels of REC deployment, in particular as they would provide the technologies, the technical and commercial skills as well as the finance needed to develop the plant serving the REC. However, these implementation models are top-down approaches where the level of citizen participation to the development, financing and decision process of the projects would likely be limited. Their social, environmental and economic benefits for the local communities interested in the project might be less easily identified. It remains to be seen whether a higher level of citizens' participation and benefits for the local communities could be achieved when the third-party facilitator is one of the ECs presented in the previous sections, i.e., initiatives themselves constituted and inspired by the objective of higher participation of the citizens to the energy transition.

Another potential avenue to guarantee higher levels of citizen participation into REC projects development could be the use of crowdfunding as a financing tool. Crowdfunding would help bottom-up initiatives to face the common barrier of access to capital, while allowing higher representation of citizens in the ownership of the project.

Provisional analysis of the financial viability of deploying a REC highlights how the boundaries given by the regulation limiting the size of a REC also affect its commercial potential. The relatively small size of the projects and the relative investment does not easily allow to amortize the fixed costs of the organization and the constitution of the legal entity responsible of the REC.[5] Thus, economies of scale might be key to RECs' commercial success. A wider and larger definition of REC in the regulatory framework would most likely favor the future expansion of the sector. In the meantime, under the current regulation, a way forward to achieve them seems to be the development of several RECs by a third party under the same legal entity, with a standardized approach. This third party could be a private sector actor, such as ESCOs, renewable developers, or utilities as well as the existing ECs presented in the previous sections.

Initial evidence from the market seems to show that such a variety of actors is entering the Italian REC market. In late 2019 RSE (a public company devoted to research on the energy system) launched a call for REC pilot projects, before the legislative framework was defined. These pilot projects were proposed by a mix of actors: small independent ESCOs, ESCOs of major energy company groups, software companies, municipalities as well as "historical" energy cooperatives and the larger ECs presented in Sect. 5.3 (in particular È nostra and Energia Positiva).

More recently several REC initiatives have been announced all over the country after the implementation of the incentive. In many cases municipalities emerge as the facilitator of the process, acting directly (through consultants) or involving energy companies or ESCOs. At least

[5] The constitution of a legal entity is not needed for Collective Self-Consumption projects.

one project has been developed by a major energy company and another one by a local energy agency in partnership with ENEA (the Italian National Agency for New Technologies, Energy and Sustainable Economic Development). Several REC initiatives have been proposed in the Alpine region by the "historical" energy cooperatives, and both WeforGreen and E' nostra (the large ECs presented in Sect. 5.3) are also active. È nostra in particular is working to develop four RECs in association with small municipalities and one with another CE developed in the 2008–2013 period.

5.7 Conclusions

The Italian community energy sector was one of the main actors in the development of the electrical system in the early 1900s. After the nationalization of the system a quasi vertical integrated monopoly was set up: the community approach was then limited to very specific local situations until the early 2000s. The liberalization of the energy markets as well as the introduction of the feed-in tariff made it possible to develop some new experiences earlier than in other European countries, although limited in number and dimension. After decades of inaction the CE sector in Italy has experienced a new growth between 2008 and 2013 with the development of initiatives aimed at citizen engagement in the energy transition. The majority were local energy community projects, mostly developing PV plants generally of a size below 100 kW, and only very few were initiatives with wider territorial scope and able to develop megawatt size plants or different projects summing up to several hundred of kilowatts. A major driver for this new wave of CE initiatives in Italy has been the implementation of the FiT support scheme, which has made PV investments quite profitable and relatively low risk, thus suitable for shared ownership projects and accessible to small scale, local projects. After the discontinuity of the FiT support in 2013 only three larger initiatives have managed to continue their activities because they have grown to a national scale, have developed multiple projects, and have expanded their member base over time, also thanks to a progressive diversification of their proposition, i.e., offering a combination of production with consumption of green electricity.

The national evolution of the regulatory framework for energy communities jointly with the renewed national support to renewable energy, implemented in July 2019, will progressively shape the CE sector in Italy, which might be on the verge of a profound evolution. As of February 2020, only a first step has been taken by the national legislators (Law 8/2020 and its regulatory implementation). It is still too early to derive any consistent conclusion on the possible development of the CE sector in Italy, but some first indications emerge. The number of initiatives announced since the full implementation of the new regulation in December 2020 is already higher than those developed in the 2008–2013 period. It is therefore possible that the community energy sector is on the verge of a development that could also lead to a number of initiatives in Italy in line with those developed in countries where this type of approach has historically been successful. It remains to be clarified how many of these initiatives under development can be actually piloted, and what multiplicative potential they will be able to feed.

This legislative and regulatory framework is provisional and it has been designed to verify potential pitfalls of the introduction of RECs in the Italian energy system. The market is still in its infancy, not just due to the fact that the policy framework is very recent, but also because its potential is currently limited by the implementation boundaries given by the regulator, which enables only small scale and local initiatives (below 200 kW).

Which other CE implementation models will be supported by the legislator will depend on the policy decisions that will be taken in the future steps of the EU Directive implementation process.

The current regulation has opened the space for the entrance in the market of third parties, such as ESCOs, renewable developers, utilities and larger players, which could act as facilitators of the development and deployment of RECs. Their possible role within the Italian EC sector will be further defined in the future evolution of the legislative framework for RECs and CECs. The entrance of large energy and institutional players might provide a boost the deployment of RECs which could be welcomed in the view of achieving the energy purpose of this regulation:

allowing higher levels of renewable generation capacity into the Italian energy system.

In terms of ECs and models of citizen participation in the development and financing of RECs (as presented in Sects. 5.2 and 5.3), whether these regulatory and market developments will lead to a revival of local, small-scale experiences as those developed in the 2008–2013 period or will reinforce the national paradigm developed by the larger Italian CE initiatives (or a combination of both) is an open question worthy of analysis and discussion in the future.

An open question remains regarding how third parties, including ESCOs, big players in the energy market as well as the existing Italian ECs, will be able to reconcile the size of their business models with the dynamics of community engagement at the local level, including the possibility of guaranteeing a high level of participation of their members in the ownership of the projects and the decision processes. In other words, whether and through which implementation model the basic principles introduced by the directives, bottom-up control and local social and environmental benefits for the communities will be ensured.

References

Bauwens, T., Gotchev, B., & Holstenkamp, L. (2016). What Drives the Development of Community Energy in Europe? The Case of Wind Power Cooperatives. *Energy Research and Social Science, 2016*(13), 136–147.

Candelise, C., & Ruggieri, G. (2020). Status and Evolution of the Community Energy Sector in Italy. *Energies, 13*(8), 1888. https://doi.org/10.3390/en13081888

Cobut, L. (2021). *Renewable Energy Communities in the European Energy Policy: An Evidence of Clean Energy for All Europeans, Not By All.* Egmont paper 111. Available at: https://www.egmontinstitute.be/renewable-energy-communities-in-the-european-energy-policy/. Accessed on 13 Apr 2021.

Confcooperative. (2021). *Confcooperative Website.* Available at: https://www.consumo.confcooperative.it/I-SETTORI/Elettrico. Accessed on 13 Apr 2021.

Creamer, E., Taylor Aiken, G., Van Veelen, B., Walker, G., & Devine Wright, P. (2019). Community Renewable Energy: What Does it Do? Walker and

Devine-Wright (2008) Ten Years On. *Energy Research and Social Science, 2019*(57), 101223.

Cringoli, G. (2017). *L'integrazione competitiva. L'industria elettrica italiana prima della nazionalizzazione*. Doctoral thesis – Dottorato in scienze storiche, archeologiche e storico-artistiche – Università degli Studi di Napoli "Federico II".

Di Gaspare, G., Dell'Oste, E.F., Williams, W. (2003). *Libro bianco su cooperazione e autoproduzione elettrica in Italia*. Ed. Città Nuova, Rome.

Energy and Strategy Group. (2020). *Le Energy Community in Italia: l'evoluzione del quadro normativo e le ricadute attese per il sistema-paese*. Available at: https://www.energystrategy.it/report/seminario-energycommunity.html. Accessed 12 Apr 2021.

Gangale, F., & Caminiti, N. M. (2005). Le politiche e le misure del settore elettrico. La liberalizzazione del mercato elettrico, le fonti rinnovabili e l'efficienza energetica. 2005 ENEA.

GSE. (2016). *Solare Fotovoltaico – Rapporto Statistico 2015*. Gestore Servizi Elettrici.

GSE. (2020). *Solare Fotovoltaico – Rapporto Statistico 2019*. Gestore Servizi Elettrici.

Hannon, M. J., & Bolton, R. (2015). UK Local Authority Engagement with the Energy Service Company (ESCo) Model: Key Characteristics, Benefits, Limitations and Considerations. *Energy Policy, 2015*(78), 198–212.

Legambiente. (2016, May). *Comuni rinnovabili 2016. La mappatura delle fonti rinnovabili nel territorio italiano*. Legambiente Report. Energy and Climate Department. Available online: https://www.legambiente.it/sites/default/files/docs/rapporto_comuni_rinnovabili_2016.pdf. Accessed on 2 Apr 2020.

Rutherford, J., & Jaglin, S. (2015). Introduction to the Special Issue—Urban Energy Governance: Local Actions, Capacities and Politics. *Energy Policy, 2015*(78), 173–178.

Scaramellini, G. (1995). *Cent'anni di luce. La società elettrica di Chiavenna 1894–1994*. Società per l'illuminazione elettrica in Chiavenna.

Spinicci, F. (2011). *Le Cooperative di Utenza in Italia e in Europa. Euricse Research Report, N.2/2011*. Available online: http://www.euricse.eu/wp-content/uploads/2015/03/1296748019_n1615.pdf. Accessed on 10 Apr 2021.

Yildiz, Ö. (2014). Financing Renewable Energy Infrastructures Via Financial Citizen Participation—The Case of Germany. *Renewable Energy, 2014*(68), 677–685.

6

Community Energy in Germany: From Technology Pioneers to Professionalisation under Uncertainty

Lars Holstenkamp

6.1 Introduction

Germany certainly belongs to the group of countries with the strongest community energy sector in Europe and worldwide. It is often presented as a community energy front-runner, together with Denmark (Morris & Jungjohann, 2016; Toke, 2011). While there seems to be widespread general support for community energy as such, at the same time, there are manifold contestations (Hein & Holstenkamp, 2018) over the role that it is supposed to play in the German "Energiewende" (energy transition) (Lienhoop, 2018). This includes the value and definition of community energy—which in German is called "Bürgerenergie" (literally:

L. Holstenkamp (✉)
Leuphana University of Lüneburg & ECOLOG Institute for Social-Ecological Research and Education, Lüneburg, Germany
e-mail: holstenkamp@uni.leuphana.de

© The Author(s), under exclusive license to Springer Nature Switzerland AG 2021

F. H. J. M. Coenen, T. Hoppe (eds.), *Renewable Energy Communities and the Low Carbon Energy Transition in Europe*, https://doi.org/10.1007/978-3-030-84440-0_6

citizen energy).[1] The recent attempt to legally define the term and introduce exemptions for wind auctions largely failed, i.e., it resulted in project developers (mis)using the rule to secure their own project pipeline through "communities" initiated and controlled by the developers themselves (Tews, 2018; see also Sect. 6.3.4). Moreover, the political focus shifted from financially participating citizens to financial gains for municipalities (Bundesministerium für Wirtschaft und Energie [BMWi], 2020).

These turns and twists in energy policy leave community energy companies in a struggle to find new business models (Herbes et al., 2017). The recent amendment of the German Renewable Energy Sources Act (Erneuerbare-Energien-Gesetz) (EEG) has not changed this situation. Despite the deadline for transposing the recast Renewable Energy Directive (RED II) into national law until mid-2021, the German legislator has not (yet) taken any step to build a definition of energy communities and further support structures into the EEG. Thus, there currently is no "revival" of community energy—in the sense of growing numbers of initiatives and installations—observable taking the Clean Energy Package (CEP) as a "window of opportunity", as may be the case in other countries with less strong community energy heritage.

In this chapter, the development of community energy in Germany is traced from rural electrification cooperatives at the end of the nineteenth and beginning of the twentieth century over renewable energy technology pioneers in the 1980s and early 1990s to more recent developments, which are characterised by uncertainties and pushes for professionalisation of the community energy sector. This historical narrative largely draws on secondary literature, including previous work by the author and data from different sources (Deutscher Genossenschafts- und Raiffeisenverband [DGRV], 2020; Kahla et al., 2017; Weiler et al., 2021), and makes use of some concepts taken from historical institutionalism (Hager & Hamagami, 2020; Stefes, 2016). Since the development of community energy—as business models of any other actor in the energy sector—depends on public regulations and market structures,

[1] Throughout the chapter, the terms "citizen energy" (as literal translation from German "Bürgerenergie"), "community energy", and "energy communities" are used interchangeably.

Sect. 6.2 gives an overview of the country background. Section 6.3 briefly highlights the historical development of the German community energy sector. Section 6.4 concludes with an outline of potential pathways.

6.2 Country Background: Brief Characterisation of Energy Markets and the Community Energy Sector in Germany

The German energy transition is characterised by an increase in the share of renewable energies, especially in the electricity sector, the nuclear phase-out and, more recently, a reduction in coal-fired power capacities. In 2020, renewable energies made up half of net electricity generation (wind energy: 27.0 per cent, solar: 10.5 per cent, biomass: 9.3 per cent, hydro power: 3.7 per cent), whereas the share of electricity from hard coal and lignite together was 24.1 per cent, nuclear power 12.5 per cent and natural gas 12.1 per cent (Fraunhofer Institute for Solar Energy Systems, 2021). In 2000, the German government negotiated with the industry to phase out nuclear energy until 2022. After some back and forth (see Table 6.1) this decision has mainly been reinstated. More recently the legislator has enacted a law to phase out coal until 2038. Within the energy transition, heating and transport tend to get less attention. The share of renewables in these sectors is much smaller than that of the electricity sector.

The "Big Four" companies E.ON, RWE, EnBW and Vattenfall still hold large shares of energy markets. However, the market concentration in electricity generation has been decreasing in recent years.

Wholesale electricity prices have dropped due to, among other factors, the increasing share of renewables. Retail prices for private households, on the other hand, are relatively high because of various levies and surcharges. They are among the highest in the EU member states. This is also why there are large incentives for prosuming on an individual household level.

The most important laws for the question at hand are the German Energy Industry Act (Energiewirtschaftsgesetz) (EnGW) and the

Table 6.1 Major developments in the German legal framework related to renewable energies and energy communities

Year	Energy (and Financial) sector developments	Renewable energy regulation (esp. EEG)	Interaction with the EU level/ other countries
Phase 2: Early community renewable energy			
1973	1st Oil Crisis		
1975	Anti-nuclear protests in Wyhl		Franco-German alliance
1979	2nd Oil Crisis		
	Three Mile Island accident		
1980	Interim report by Enquête Commission (1979–83) indicates change away from nuclear		"Soft energy path" (Lovins, 1977)
	"Energiewende" book by Öko-Institut (Krause et al., 1980)		
1983		(Failed) Growian project (large-scale windmill)	
1986	Chernobyl disaster		
1990		Electricity Feed-In Act (StrEG) enacted	Role models: Danish FiT, US PURPA of 1978
1991		StrEG into force	
1993	Energy consensus talks		
1994/5		"Aachen model" of cost-covering remuneration implemented in Aachen, Freising, Hammelburg	
1996		Privileging of wind turbines in outdoor areas, Federal Building Code (BauGB), Sec. 35	VDEW complaint with DG Competition
1997	Merger: Badenwerk & EVS form EnBW		
Phase 3: Mainstreaming			

1998	Amendment of Energy Industry Act (EnWG): liberalisation of energy markets	StrEG 1998: new compensation mechanism	PreußenElektra files lawsuit against StrEG 1998
1999		100,000 rooftop programme	
2000	Mergers: RWE & VEW; VIAG & VEBA form E.ON "Nuclear Consensus"	Renewable Energy Sources Act (EEG) enacted and into force (EEG 2000): fixed FiT for 20 years, feed-in priority	
2001			EU Court of Justice: *PreußenElektra v Schleswag* ⇒ StrEG no state aid
2002	Merger: Vattenfall Europe formed ⇒ *"Big Four"*		
2003		Special Equalisation Scheme Act	
2004		EEG 2004 enacted and into force Cost-efficiency concerns BMWi: Tendering	
2006	Amendment of German Cooperative Societies Act (GenG)		
2008		EEG 2009 enacted	
2009		EEG 2009 into force	
2010	Extension of operational lifespan of nuclear power plants ["Laufzeitverlängerung"]		
2011	Fukushima accident ⇒ "Ethics Commission", reinstalled phase-out Big Four in economic crisis	EEG 2012 enacted Discussion about "market integration" & distributional effects	

(*continued*)

Table 6.1 (continued)

Year	Energy (and Financial) sector developments	Renewable energy regulation (esp. EEG)	Interaction with the EU level/other countries
2012		EEG 2012 into force: market premium (FiP) "PV amendment" of June 2012: tariff reductions, solar cap (52 GW), corridor for expansion, "market integration model"	
Phase 4: Stagnation & recent developments			
2013	German Investment Code (KAGB) enacted Responsibility for energy transferred from BMU to BMWi	BMWi proposal on "electricity price break"	Commission opens state aid investigation
2014		EEG 2014 enacted and into force: trial out of tenders, "expansion paths" (caps) for all renewables	Environmental Protection and Energy Aid Guidelines ⇒ auctions as "default" Commission decision: EEG 2012 state aid
2016		EEG 2017 enacted	DE government appeals v decision of General Court Clean Energy Package (CEP) proposed
2017		EEG 2017 into force: tenders for >750 kW, legal definition of community energy company [Bürgerenergiegesellschaft]	

2019	"Coal Commission" ⇒ "Coal Consensus" = phase-out until 2038		EU Court of Justice: EEG 2012 no state aid
2020		EEG 2021 enacted	EEG notified as state aid
2021		EEG 2021 into force	Deadline for transposition of RED II

Source: Own summary, partly based on (Jacobsson & Lauber, 2006; Morris & Jungjohann, 2016; Stefes, 2016; Toke, 2011)

EEG. Further support schemes beyond those regulated with the EEG include low interest loans provided by the national development bank KfW and Rentenbank, Germany's development agency for agribusiness. KfW and the Federal Office for Economic Affairs and Export Control (Bundesamt für Wirtschaft und Ausfuhrkontrolle) (BAFA) administer support programmes in the heating sector, e.g., for (biomass- or biogas-fired) district heating grids.

In the renewable energy sector, citizens own a large proportion of the installed capacity, even though their share has been declining over recent years. Starting with small pioneering groups, there typically have been initiatives and social networks, partly from the environmental movement—in the following called "community energy initiatives"—behind legal owners of renewable energy installations—in the following called "community energy companies". Kahla et al. (2017) estimate the number of community energy companies in the range of 2000 to 2500, counting separately those SPVs which have been established by the same group of citizens. Overall, the German community energy sector is characterised by its heterogeneity in several respects, including initiators and supporting structures (intermediaries, social networks), geographical scale (local, regional, national; membership, activities), size (members, investment volumes), technologies, involvement of local authorities, legal forms and forms of financing. While municipalities engage with community energy initiatives and companies—Creamer et al. (2018) call this "entanglements"—direct ownership by municipalities and municipally-owned companies is usually not considered part of the German "Bürgerenergie" (Morris & Jungjohann, 2016). Moreover, some community energy companies can be characterised as (also) profit-oriented. The profit motive may overtake other motivations in some cases or for parts of the members (Holstenkamp & Kahla, 2016; similarly in other countries such as Austria: Fleiß et al., 2017).

Common legal structures in the community energy sector include registered cooperatives and limited liability partnerships, for small projects and also civil law associations. Other legal structures such as joint-stock companies or limited liability companies are less common (Holstenkamp et al., 2018; Yildiz et al., 2019). Two kinds of legal structures can be differentiated: First, the creation of single special purpose vehicles (SPVs) per project, i.e., legal entities that hold the installations as assets on their balance sheets, and second, the formation of a community energy company that holds several projects and develops over time. The former is typical of the wind energy sector and large solar PV projects. Registered cooperatives usually belong to the latter case.

Community energy investments in a wider sense include different forms of financing such as bearer bonds, sub-ordinated loans or savings certificates and until some years ago also participation rights or certificates (Holstenkamp, 2014, 2015a; Yildiz et al., 2019). Roughly one-fifth of municipal utility companies, which exist in large number—ca. 1000—in Germany, offer such investment opportunities to local inhabitants or their customers, typically together with a local bank. While some municipal utility companies regard community energy as a competitor, others cooperate with these groups or even (help to) establish community energy companies. Similarly, some citizens engaged in community energy initiatives see municipal utilities as part of the "old electricity sector" that they want to replace (Schreuer & Weismeier-Sammer, 2010), while others actively search for cooperation and support.

6.3 The Historical Development of Energy Communities and Their Role in the German Energy Market

6.3.1 Fast Forward Through the Historical Development and Preliminary Remarks on Periodisation

In the following, the historical development of the community energy sector will be described in several periods from early electrification until today. Periodisation, i.e., the identification of historical phases, is always challenging (Jordheim, 2012; Lieberman, 2001; Lorenz, 2017). Leiren and Reimer (2018), for instance, organise their historical account along the versions of the EEG. Jacobsson and Lauber (2006), on the other hand, divide the history of renewable energy policy in Germany into a formative period (1974–1988), take-off of wind, but not for PV (1988–1998) and take-off for solar, continued growth of wind energy and new political challenges (1998–2003). Oriented at the rise, fall and renaissance of community approaches in the German energy sector, this section is divided according to four main phases: First, early experiences of cooperative approaches in the electricity sector, i.e., with the establishment of the electricity industry at the end of the nineteenth century until the beginning of concentration in the German energy sector. While some, but probably relatively few, cooperatives were established in the wake of the oil crises, the appearance of renewable energy pioneers is considered to mark the beginning of the second phase. The third phase can be characterised as "mainstreaming", which is closely related to the introduction of the EEG and later the rise of the community energy cooperative. The stagnation of new foundations of this type around 2013 forms the turning point to the recent past and present that are described in Sect. 6.3.4.

While there is no comprehensive collection of data for the whole community energy sector so far, several databases exist, which can be used to describe developments in community ownership (Fischer & Wetzel, 2018; Kahla et al., 2017; Weiler et al., 2019a). From these numbers, it is possible to delineate the stylised phases and (ideal) types of community energy in Germany across time as depicted in Fig. 6.1.

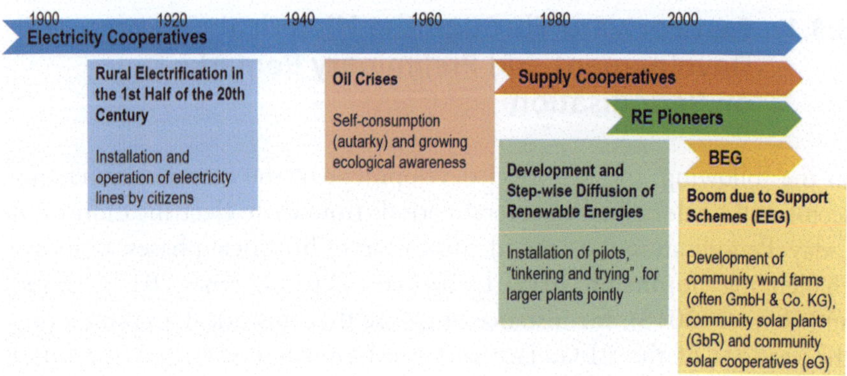

Fig. 6.1 Types of community energy in Germany across time

6.3.2 Early Cooperative Experiences in the Electricity Sector

Cooperative approaches in the energy sector are not a fundamentally new concept. Electricity cooperatives have existed in several countries, especially to electrify rural areas. Engaged citizens often acted together with local authorities. Countries include, among others, Spain (Cuesta-Fernandez et al., 2020), the USA since Roosevelt's New Deal (Pellegrini & Tasciotti, 2012; Wolman, 2007), Italy (Mori, 2013) or Switzerland (Gugerli, 1996; Schmid et al., 2020). At its peak around the late 1930s/early 1940s, there had been more than 6000 electricity cooperatives in the former German Reich (1871–1945), especially in rural areas (Holstenkamp, 2015b, 2018). In these regions, the cooperatives together had a significant market share. At the same time, the cooperative electricity sector was relatively heterogeneous, depending highly on state level or provincial regulations and energy policy. Electricity cooperatives mainly addressed market and government failures in electrifying remote rural areas.

Numbers decreased substantially due to the politically enforced, but also partly economically caused, concentration process in the German energy sector (Hughes, 1983; Stier, 1999). In several states, mergers were politically enforced. In some cases public or private utility companies, which supplied the electricity distributed by electricity cooperatives, used their power and purchased cooperatives or their assets. Some electricity cooperatives, especially those established to provide for electric lighting in small districts or rural settlements, were too small to survive in the long term (Holstenkamp, 2018).

Only a few of the old cooperatives have survived until today, especially in one state, namely Bavaria (Holstenkamp, 2018). Most are relatively small companies; some only own distribution grids, the operation of which they contract out to other utility companies. However, there are also some larger cooperatives such as Alb-Elektrizitätswerke Geislingen-Steige eG (balance-sheet total of the corporate group: € 255 m), Unterfränkische Überlandzentrale eG (€ 162 m) or Teutoburger Energie Netzwerk eG (€ 53 m). Overall, electricity cooperatives have taken a similar position in Germany's centralised electricity pre-liberalisation as have municipal utility companies. In several cases, municipalities were (and are), in fact, members of local or regional electricity cooperatives.

As highlighted elsewhere, a historical analysis of economic and political interests and the economic views in the light of technological possibilities of the time can enable and enrich the thinking of scenarios (Holstenkamp & Stier, 2018). In this sense, the experiences and developments of old electricity cooperatives may act as stimulating examples to study the specific constellations of interests, experiences, and power structures that have led to the current state and scenarios for future pathways. A few hints in this direction will be provided in the concluding section.

6.3.3 Early Community Renewable Energy

Besides some new foundations in the aftermath of the 1973 and 1979 oil crises, the next rise of cooperative endeavours in the energy sector took place in the late 1980s and early 1990s. Using the Danish wind energy cooperatives as a role model, local people established several community wind groups, but also joint ventures for solar installations. Farmers experimented with biogas power plants (Bauwens et al., 2016; Mautz et al., 2008). Critical moments or "junctures" (Capoccia, 2015; Karapin, 2020) kickstarting or reinforcing (Stefes, 2020) the development of these niches, were anti-nuclear protests such as in Whyl in south-western Germany, near the French border, the oil crises and nuclear accidents, most of all the Chernobyl disaster in 1986 (see, also for the following, Table 6.1).

Probably the first community-owned wind power plant was installed in sight of the Brokdorf nuclear power plant in 1989 by uww Windstrom Wedel GmbH & Co. KG, a limited liability partnership created by members of the Hamburg-based association Umschalten e.V. with the aim to generate an alternative to nuclear energy as a reaction to the Chernobyl disaster (Byzio et al., 2002). Other early examples include community wind farms in Friedrich-Wilhelm-Lübke-Koog and Reußenköge in North Frisia, a region in the north-west of Schleswig-Holstein (Krauss, 2010; Süsser et al., 2017; Süsser & Kannen, 2017). The growing number of special purpose companies created to install wind power was mainly fuelled by a support scheme called "100 MW wind programme" of 1989, renamed the "250 MW wind programme" after extension in 1991, and by the introduction of the Electricity Feed-in Act (Stromeinspeisegesetz, StrEG) on the federal level the same year. Both—together with favourable financing conditions provided by Deutsche Ausgleichsbank (since 2003: KfW Mittelstandsbank) and KfW—made investments at some locations economically attractive, which, in turn, spurred the inflow of capital from better-off non-regional, "foreign" investors. Already for the early or mid-1990s, Rübsamen et al. (1995) observed a trend towards investments which were more motivated by profits or potential tax savings and not so much by contributing to the energy transition. In this respect, they highlight differences to solar PV and combined heat and power (CHP), but also stress the large heterogeneity of the sector. Many community wind parks were legally structured as limited partnerships with a limited liability company as general partner and manager of the project (Enzensberger et al., 2003a). As this has been a structure common in the investment fund industry in Germany, it translated easily into professionalised structures with high net worth individuals (HNWI) (Enzensberger et al., 2003a, b; Mautz et al., 2008), especially in the eastern parts of Germany with a lower number of community energy companies (Yildiz et al., 2015).

As solar PV took off only later (Jacobsson & Lauber, 2006; Stefes, 2016), projects were often promoted by small local groups. Solar-Förderverein (SFV), an association for the support of solar energy founded in 1986 with seat in Aachen, developed and advocated for cost recovering remuneration, which has been called the "Aachen model" since its introduction in this city, even though councils in Hammelburg and Freising already earlier decided to implement this type of feed-in tariff. This local support helped to bridge a gap left by the end of the "1,000 roofs programme" on the federal level (Jacobsson & Lauber, 2006)—illustrating some possibilities of the local level in Germany's multilevel governance as well as the cooperation on the local level, e.g., between solar activists, local authorities, and municipal utility companies, and the build-up of advocacy coalitions (Sabatier, 1987).

6.3.4 From Pioneers to "Mainstreaming"

As indicated above, early "mainstreaming" probably set in already with the professionalisation in the wind power sector in the mid- to late-1990s. This development was further strengthened and extended to other sectors with the Renewable Energy Sources Act coming into effect in 2000. The new Social-Democrat/Green coalition government had just established a "100,000 rooftop programme" and, in parallel, negotiated the nuclear phase-out, which was enacted by law two years later. The number of community energy companies increased steadily, most of them being legally structured as limited liability companies or civil law associations. Only a few community groups formed registered cooperatives. Moreover, old electricity cooperatives faced another phase of dissolutions following growing legal requirements and economic stress in the wake of energy market liberalisation since 1998.

This situation changed with the amendment of the Cooperative Societies Act in 2006, and with the changing image of cooperatives, increasingly seen as a modern form and suitable legal structure for an "alternative economy" (Klagge & Meister, 2018). Partly supported by marketing campaigns of some cooperative associations and activities of cooperative banks, the number of energy cooperatives grew significantly between 2008 and 2012/13 (Fischer & Wetzel, 2018; Hall et al., 2018; Kahla et al., 2017). The Fukushima disaster in 2011 further increased public awareness for alternative forms of energy production. Since then, the term "community energy cooperatives" (Bürgerenergiegenossenschaften) is used as a synonym of community energy companies and prototypical model in political debates. Most of these cooperatives were and are engaged in solar PV, often on top of public buildings using the feed-in tariff. Some more innovative cooperatives also tried out third-party financing models for on-site consumption of electricity. Similar to earlier developments in the community wind sector reported above, relatively favourable economic conditions for PV installations under the Renewable Energy Sources Act of 2009 also attracted more profit-oriented investors to become members of community energy cooperatives (Holstenkamp & Kahla, 2016).

In the wind energy sector, characterised by growth in installed capacity per windmill and, accordingly, in investment volumes. Those active in this sector often either buy shares in limited liability partnerships or create special purpose vehicles themselves to mitigate financial and legal risks (Holstenkamp & Degenhart, 2014).

Biogas-fuelled district heating cooperatives constituted the second most common type of energy cooperative (Kahla et al., 2017). They form part of a section of community energy called "Bioenergiedörfer" (bioenergy villages; Jenssen et al., 2014; Roesler, 2019; Wüste & Schmuck, 2012). Financing is based on different sources, including equity from members, connection fees, subsidies through BAFA, and/or low-interest loans from KfW. Typically these cooperatives are not geared towards generating profits, but rather to offer the lowest possible prices for heat. In that they follow a more traditional cooperative model (Mori, 2013).

6.3.5 More Recent Developments (1): Competitive Environment and Stagnation

Legal changes in 2012–2014 and the shift of responsibilities for renewable energies from the Federal Ministry for the Environment, Nature Protection, and Nuclear Safety (Bundesministerium für Umwelt, Naturschutz und nukleare Sicherheit, BMU) to the Federal Ministry for Economic Affairs (Bundesministerium für Wirtschaft, BMWi) mark a turning point with a subsequent slump in the number of new foundations of energy cooperatives. Morris and Jungjohann (2016) summarise this as a changing balance between the "New Players camp", including community energy advocates, that lost ground, and the "Polluter Pays camp", i.e., proponents of neoclassical economic approaches and instruments, both constituting competing advocacy coalitions. Intensified debates on costs, distributional effects, and the need to orient policy towards "market-based instruments" were followed by severe cuts in feed-in tariffs for solar PV in 2012, the introduction of a first voluntary and later mandatory direct marketing for larger projects, i.e., a gradual switch to a feed-in premium model with a variable premium. In 2014, the legislator also introduced pilot tenders for ground-mounted solar PV (on this policy shift: Leiren & Reimer, 2018). With the amendment of the EEG in 2017, the legislator introduced mandatory auctions for large PV and wind energy installations. The results of these auctions show that community energy has only a minor market share (Grashof, 2019; Weiler et al., 2019b, 2020a, b). While ownership may change to communities after completion of the installations in some cases, this general pattern will most probably still hold with regard to wind farms or large solar PV installations during the operational phase. Overall, auctions seem to have impacted community ownership especially in the wind onshore sector. Community ownership has also before been marginal for utility-scale PV; but auctions may deter community actors from entering this market (Weiler et al., 2021). Overall, this has contributed to the fact that the community energy sector has rather stagnated (DGRV, 2020; Fischer & Wetzel, 2018; Kahla et al., 2017).

In 2017, citizen wind companies as legally defined were able to offer bids at earlier stages of project development (Grashof et al., 2019). The legal definition allowed for a group of ten citizens living in the district in which the wind farm is installed to be called a "citizen energy company" (see Table 6.2 for more details). Several project developers used this rule to set up such citizen groups. While the implementation differed from developer to developer (Grashof et al., 2019), it can be assumed that quite a large number of these projects, if successfully completed at all, will be driven by these project developers and do not constitute community wind companies in the original sense (Creamer et al., 2019; Walker & Devine-Wright, 2008).

The legal definition in EEG 2017, Sec. 3 No. 15, deviates in several respects from energy community definitions in RED II and ED. Most prominently, energy communities could principally be composed of local authorities or (local) small- and medium-sized companies (SMEs) only, whereas in the German context—with its long-standing history of municipal utility companies that had become part of the old centralised energy sector, i.e., are "incumbents"—there is a focus on (local) citizens (Morris & Jungjohann, 2016). This is the case despite municipal utility companies engaging with community energy (Schmid et al., 2020) and legal requirements to offer at least 10 per cent of shares of community wind farms to local municipalities (EEG 2017, Sec. 36 g). Hence, sometimes "entanglements" (Creamer et al., 2018) are acknowledged in public discourses, sometimes community energy and municipal ownership are

Table 6.2 Comparison of definitions of community energy/energy communities in Germany

Source:	BWE	Monitoring Project "Diversity of Actors" Weiler, Weber, Grashof, et al. (2019)		EEG 2017, Sec. 3 No. 15	Clean Energy Package RED II, Art. 2 No. 16	Clean Energy Package ED, Art. 2 No. 11
Term	Bürgerwindpark [community wind farm]	Beteiligungsoffene Bürgerenergie [open & participatory community energy]	Beteiligungsoffene nationale Akteure [open & participatory national actors]	Bürgerenergiegesellschaft [community energy company]	Renewable Energy Community [Erneuerbare-Energien-Gemeinschaft]	Citizen Energy Community [Bürgerenergiegemeinschaft]
Participants	Citizens, companies	Citizens		Citizens 10% offer to municipalities for community wind farms (Sec. 36 g Para. 1 No. 2)	Citizens, SMEs, local authorities (incl. municipalities)	
Regionality	Village or close neighbourhood (citizens) Headquarters in municipality (companies), regionally anchored management	Main residence or seat of SPV and GP in region of facility Main activity in region of facility Region with common socio-cultural identity	*At least one of the "regionality criteria" not fulfilled*	City or county of facility, main residence (§§ 21, 22 BMG) for at least 1 year	Proximity of facility	*No regionality criterion*

(*continued*)

Table 6.2 (continued)

				Clean Energy Package	
Source:	BWE	Monitoring Project "Diversity of Actors" Weiler, Weber, Grashof, et al. (2019)	EEG 2017, Sec. 3 No. 15	RED II, Art. 2 No. 16	ED, Art. 2 No. 11
Openness	Every local inhabitant can participate Relatively low minimum invest Public offer No exclusion without good cause	Public offer (alternative criterion: Minimum of 20 natural persons) Minimum invest below thresholds: Lower threshold: ≤1000€ (East) or ≤2000€ (West), respectively Upper threshold: <5000€ (all)	Minimum of 10 natural persons	Open & voluntary participation	
Governance	Quorum (citizens): ≥70% Supervisory board made up of local residents AGM as supreme body No single person dominating (max 15% voting rights) No special rights for single persons All relevant contract transparent	Quorum (citizens): ≥51% No single person dominating (<25% voting rights/shares)	Quorum (citizens): ≥51% No single person dominating (max. 10% voting rights)	Autonomous Effectively controlled by shareholders/members	

Activity	Wind	Monitoring: Large PV (>750 kWp), wind	Special rules for wind onshore only	Renewable energies	Generation, distribution, supply, consumption, aggregation, energy storage, energy efficiency services, charging services for EVs, other energy services to members or shareholders
Others	Commercial tax remains predominantly in municipality			Primary purpose: Environmental, economic or social community benefits rather than financial profits	

juxtaposed, especially where the latter is considered part of the "old fossil- and nuclear-based electricity sector" by environmental activists or, more recently, in discussions on distributional effects. As definitions by the German Wind Energy Association (Bundesverband WindEnergie, BWE) and Weiler et al. (2019a) illustrate, there is also no common agreement in Germany, at least in details, as to what constitutes community energy or "Bürgerenergie" (see Table 6.2). This, again, reflects the heterogeneity of the German community energy sector and the difficulties of demarcating it from other types of ownership.

While these policy shifts certainly have had the most tremendous impact on the community energy sector, the introduction of the German Investment Code (Kapitalanlagegesetzbuch, KAGB) in 2013 at least temporarily hindered developments due to high uncertainties around the interpretation of several provisions (Holstenkamp, 2014). This put additional stress especially on those cooperatives which wanted to re-orient themselves from solar PV towards an engagement in wind energy projects as a reaction to feed-in tariff cuts for PV.

6.3.6 More Recent Developments (2): Focus on 'diversity of actors' and State-Level Activities

However, there is also another facet in recent policy discourses in Germany: Since 2014, there has been an explicit actor-related target in the Renewable Energy Sources Act: "maintain" "stakeholder diversity" (EEG 2017 Sec. 2 para. 3 sentence 2) or "retain" "diversity of players" (EEG 2014 Sec. 2 para. 5 sentence 3).[2] The ministerial draft of the 2014 act still contained the term "citizen participation"; this was subsequently changed into "diversity of actors" in the legislative procedure. Other documents include the term "small actors" as the addressee of public support. Maintaining a "diversity of actors" is certainly not a goal in itself, but rather a means to achieve other goals, especially social project-related

[2] The original German term is in both bases "Akteursvielfalt (...) erhalten"—maintain or preserve the diversity of actors.

acceptance (Weiler et al., 2019a). Moreover, the idea is not to preserve a certain market structure or shares for specific types of actors, but rather to ensure that a certain degree of diversity is maintained. Nevertheless, in policy documents community energy companies, especially energy cooperatives, are often mentioned. Overall, the concrete meaning of the term stays vague. In general, the idea is to maintain a certain level of heterogeneity in ownership of renewable energy assets. This may or may not include community energy or energy communities as defined in RED II and ED.

Explicit support for community energy beyond the special treatment of community wind farms in EEG 2017, Sec. 36 g, is currently taking place at state (see Table 6.3) and municipal levels (Schmid et al., 2020). In the state of Mecklenburg-Western Pomerania, discussions around

Table 6.3 Selected community energy-related activities in the German federal states

Baden-Württemberg	EnBW Regional AG, subsidiary of state-owned utility company EnBW, supports foundation of association of community energy cooperatives in Baden-Württemberg (VBBW)	2009
	State network of volunteer energy initiatives (LEE)	
Hesse	Economic activity by municipalities in the energy sector: Citizen participation shall be enabled (Hessian Municipal Code, Sec. 121, para. 1a sentence 2)	2012 2013
	Support of energy cooperative network LaNEG Hessen	
Mecklenburg-Western Pomerania	Legislation enhancing the participation of citizens and municipalities in wind energy projects (BüGemBeteilG MV), *legally contested*	2016
North Rhine-Westphalia	Platform community energy and energy cooperatives, hosted by state energy agency (Energieagentur.NRW)	2016
Rhineland-Palatia	Support of exchange and energy cooperative network LaNEG	2011/2
Schleswig-Holstein	Citizen Energy Fund: *advice & risk capital*	2018
	Similar instrument recommended for federal government (BT-Ds. 212/20)	2020
	Guidelines for the assessment of fair wind farm planners in Schleswig-Holstein, *includes guidelines on financial participation*	2018

wind energy development and social acceptance led to the introduction of mandatory financial participation of municipalities and local citizens in 2016, a law built after the Danish model (Olsen, 2014). In some regions such as North Frisia, other legal instruments are used to favour community-owned wind (Maly, 2014).

The EEG was amended in late 2020. While former rules for community energy have been kept, there seems to be a tendency to exclude any support for community energy groups from the EEG and search for other support schemes such as special loans or risk capital. The state of Schleswig-Holstein has set up such a fund for supporting the early development phase of community energy projects and the state government also proposes such an instrument on the national level. The German Environment Agency is working on such a proposal.

In the meantime, the German government has submitted its National Energy and Climate Plan (NECP) for 2030. While the draft plan did not mention renewable energy communities (RECs) or citizen energy communities (CECs), assuming that there is no need to take further measures to create a level playing field with incumbent and larger energy companies, the government does not mention renewable energy communities or "Bürgerenergie" often in its final NECP either. In this respect, the German NECP mainly contains a short sentence on energy communities stating that it will be checked if any action is necessary to thoroughly transpose the CEP directives into national laws. However, the amended EEG does not include any definition of energy communities. There are no targets for energy communities. Moreover, the landlord-to-tenant electricity model is the only regulation for forms of collective self-consumption (Campos et al., 2020). The rather complex rules around the landlord-to-tenant electricity model are an example of path dependence and layering as changing regulations tend to build on this established concept. The German parliament, the Bundestag, has passed a resolution which contains a review mandate, namely to investigate whether the rules for self-generated electricity can also be transferred to RECs. The 2021 version of the EEG includes rules for financial participation of municipalities, though on a voluntary basis and not mandatory—marking a shift of attention from direct citizen participation to municipalities, somewhat similar to the shift from community energy to local energy in the UK (Devine-Wright, 2019).

As a reaction to changes in the legal environment, many community energy initiatives look for new business models (Ehrtmann et al., 2021; Herbes et al., 2017). Overall, community energy business models are becoming more diverse (Kahla et al., 2017).

6.4 Future Prospects

So far, it is unclear if the Clean Energy Package will have any effect on the community energy sector in Germany. Overall, Germany traditionally has a strong community energy sector and, relative to other countries, a strong regulatory framework for collective forms of prosumerism (Campos et al., 2020). However, there are still several provisions in the CEP that have not (yet) been transposed to national law with the amendment of the Renewable Energy Sources Act at the end of 2020. The major argument seems to be that it is not anymore necessary to create a level playing field for community energy in Germany vis-à-vis other market participants. While there is a large general consensus to preserve the existing diversity of actors and especially support community energy initiatives, there seem to be different understandings of the concrete steps to be taken and the ways how to assist the community energy sector. Much will probably depend on the results of federal elections in autumn 2021 and the following coalition talks.

At the same time, there are several initiatives on state level to support community groups (see Table 6.3). As the history of community solar illustrates, initiatives even at municipal level may also help energy communities to develop. While national level changes may develop a greater leverage effect, dynamics in the multilevel governance system including state and municipal levels should be taken into account.

To summarize, the CEP has not been used (yet) as a window of opportunity to foster or revitalise Germany's stagnating community energy sector—despite or because of being a community energy front-runner.

Because of being a front-runner as an example of the "law of the disadvantageous lead" (Coenen and Hoppe, in the introductory and concluding sections of this book) or a first-mover disadvantage, because of the following reasons:

1. A well-developed sector means that new regulations have to fit to existing types of ownership and to legal or social concepts available ("institutional fit"), which may be difficult as the discussion around the legal definition of "Bürgerenergiegesellschaften" in Germany illustrates. Even within the "New players camp", as Morris and Jungjohann (2016) call it, there are different variations and motivations, with a recent shift of focus from citizen to municipal ownership and financial participation of municipalities, since the latter is considered more inclusive. This juxtaposition of both types, despite of its entanglements in reality and the inclusion of both in the EU's energy community concepts, is an example of how path dependence plays out in this respect.
2. The German government can always point to the existence and size of the community energy sector, which still is—also in relative terms—larger than in many other EU member states. As such, the national government may interpret regulations as addressing latecomer countries rather than Germany.

This said, there are several other elements to be considered—as arguments rather for a despite of Germany being a community energy front-runner:

1. Timing: In principle, CEP & transposition could be work out as "critical junctures" even in a front-runner country. The REC and CEC concepts have to be transposed into national laws after they have been established at the EU level, i.e., and a higher level in the multilevel governance system. In this sense, the EU level constitutes just another "action arena" with rules and actors slightly different from the national arena. Nevertheless, the introduction of the CEP coincided with EEG 2017 discussions around the legal definition of community energy and the failed attempt to protect the intended target group with special rules for community wind farms. At a different time, it may have worked out—or will in the future.
2. Sub-national and multilevel policy dynamics: It is most likely that the national legislator will follow the example of some federal states and introduce some support measures for community energy, but outside

of and not within the EEG. The EU has not directly defined the type of instruments to be included in a supporting framework for energy communities. Hence, there is some space for national governments to decide on the way to legally implement it. In this sense, the CEP may play out as another "layer" developed in multilevel policy over time rather than initiating a radical change ("path creation"). Moreover, the enactment of RED II and ED took place only a few years ago and transposition is still ongoing. Hence, it may be too early for judgements considering more complex dynamic policy processes between several levels.

3. Dominating narrative: While the EU has introduced prosumers and energy communities through the CEP, in German national energy policy a different narrative first competed and since around 2012/13 started to dominate, emphasising "market integration" and allocative efficiency, including auctions as instruments and creating "level playing fields" rather than supporting specific types of actors or businesses. Historical institutionalists have described this as an incremental change through "institutional layering", which also affects the community energy sector (Hager & Hamagami, 2020).

For the individual community energy companies, especially multi-project energy cooperatives, there are probably four main strategic options:

- "Stagnation": Community groups keep the assets that they operate, but do not engage in further projects.
- "Sale": The community energy company sells its assets or its members their shares.
- "Merger" and/or "cooperation": Two or more cooperatives merge and build a larger entity, which holds more assets and has more economic strength to also employ personnel (see next point). A weaker form would be cooperation.
- "Growth & professionalisation": Growth may also allow to hire staff or pay for previously voluntary work.

Among these strategies, professionalisation has received some attention in the literature (Brummer & Herbes, 2018; Herbes et al., 2017;

Horstink et al., 2020; Nolden et al., 2020). While reliance on volunteers creates a certain degree of "fragility" that can impede further growth and development of community energy (Brummer, 2018), professionalisation may also have some flipsides such as the danger of "mission drift" (Rothschild et al., 2016), less engagement by volunteers—according to Rübsamen et al. (1995) probably especially women—and less participatory decision-making (Brummer & Herbes, 2018), which, in turn, can have an impact on production of social capital (Cumming, 2008; Markowitz & Tice, 2002). Moreover, it may make energy cooperatives become somewhat like municipal utility companies in a process of "institutional isomorphism" (DiMaggio & Powell, 1983).

Mergers can already be observed in the cooperative energy sector. Moreover, there are several "secondary" cooperatives or companies legally structured in a different form, which are owned by "primary" cooperatives (and community energy companies of other legal structure). Such a type of "joint cooperative action" (Bauwens et al., 2016) is prominently exemplified by the case of Bürgerwerke eG in the field of electricity retailing (Ehrtmann et al., 2021). The electricity retailing business also illustrates the need for cooperation to open up new areas of business and to develop new business models—a task that according to Nolden et al. (2020) specialised intermediaries may take over. Most likely we will see all four strategies employed by different community energy players. Moreover, there have been somewhat similar strategies and developments among old electricity cooperatives—something to be further investigated in future work.

References

Bauwens, T., Gotchev, B., & Holstenkamp, L. (2016). What Drives the Development of Community Energy in Europe?: The Case of Wind Power Cooperatives. *Energy Research & Social Science, 13*, 136–147. https://doi.org/10.1016/j.erss.2015.12.016

Brummer, V. (2018). Community Energy—Benefits and Barriers: A Comparative Literature Review of Community Energy in the UK, Germany and the USA, the Benefits it Provides for Society and the Barriers it Faces. *Renewable and Sustainable Energy Reviews, 94*, 187–196. https://doi.org/10.1016/j.rser.2018.06.013

Brummer, V., & Herbes, C. (2018). Of expertise, Social Capital, and Democracy: Assessing the Organizational Governance and Decision-Making in German Renewable Energy Cooperatives. *Energy Research & Social Science, 37*, 111–121. https://doi.org/10.1016/j.erss.2017.09.03

Bundesministerium für Wirtschaft und Energie. (2020). *Finanzielle Beteiligung von Kommunen und Bürgern am Betrieb von Windenergieanlagen: Eckpunktepapier des BMWi.* Retrieved from https://www.erneuerbare-energien.de/EE/Redaktion/DE/Downloads/eckpunktepapier-finanzielle-beteiligung.pdf?__blob=publicationFile&v=2

Byzio, A., Heine, H., & Mautz, R. (2002). *Zwischen Solidarhandeln und Marktorientierung: Ökologische Innovation in selbstorganisierten Projekten – autofreies Wohnen, Car Sharing und Windenergienutzung. SOFI-Berichte.* SOFI. Retrieved from http://sofi-goettingen.de/fileadmin/SOFI-Graue-Reihe/Zwischen_Solidarhandeln_und_Marktorientierung.pdf

Campos, I., Pontes Luz, G., Marín-González, E., Gährs, S., Hall, S., & Holstenkamp, L. (2020). Regulatory Challenges and Opportunities for Collective Renewable Energy Prosumers in the EU. *Energy Policy, 138*, 111212. https://doi.org/10.1016/j.enpol.2019.111212

Capoccia, G. (2015). Critical Junctures and Institutional Change. In J. Mahoney & K. A. Thelen (Eds.), *Strategies for Social Inquiry. Advances in Comparative-Historical Analysis* (pp. 147–179). Cambridge University Press.

Creamer, E., Aiken, G. T., van Veelen, B., Walker, G., & Devine-Wright, P. (2019). Community Renewable Energy: What Does It Do?: Walker and Devine-Wright (2008) Ten Years on. *Energy Research & Social Science, 57*, 101223. https://doi.org/10.1016/j.erss.2019.101223

Creamer, E., Eadson, W., van Veelen, B., Pinker, A., Tingey, M., Braunholtz-Speight, T., & Lacey-Barnacle, M. (2018). Community Energy: Entanglements of Community, State, and Private Sector. *Geography Compass, 12*(7), e12378. https://doi.org/10.1111/gec3.12378

Cuesta-Fernandez, I., Belda-Miquel, S., & Calabuig Tormo, C. (2020). Challengers in Energy Transitions Beyond Renewable Energy Cooperatives: Community-Owned Electricity Distribution Cooperatives in Spain. *Innovation: The European Journal of Social Science Research, 33*(2), 140–159. https://doi.org/10.1080/13511610.2020.1732197

Cumming, G. D. (2008). French NGOs in the Global Era: Professionalization "Without Borders"? *Voluntas: International Journal of Voluntary and Nonprofit Organizations, 19*(4), 372–394. https://doi.org/10.1007/s11266-008-9069-5

Deutscher Genossenschafts- und Raiffeisenverband. (2020). *Energiegenossenschaften 2020: Jahresumfrage des DGRV*. Retrieved from https://www.dgrv.de/wp-content/uploads/2020/07/20200701_DGRV_Umfrage_Energiegenossenschaften_2020-1.pdf

Devine-Wright, P. (2019). Community versus Local Energy in a Context of Climate Emergency. *Nature Energy, 4*(11), 894–896. https://doi.org/10.1038/s41560-019-0459-2

DiMaggio, P. J., & Powell, W. W. (1983). The Iron Cage Revisited: Institutional Isomorphism and Collective Rationality in Organizational Fields. *American Sociological Review*, 147–160. https://doi.org/10.2307/2095101

Ehrtmann, M., Holstenkamp, L., & Becker, T. (2021). Regional Electricity Models for Community Energy in Germany: The Role of Governance Structures. *Sustainability, 13*(4), 2241. https://doi.org/10.3390/su13042241

Enzensberger, N., Fichtner, W., & Rentz, O. (2003a). Evolution of Local Citizen Participation Schemes in the German Wind Market. *International Journal of Global Energy Issues, 20*(2), 191–207. Retrieved from http://inderscience.metapress.com/content/UR4LG6D2CRXTKQX3

Enzensberger, N., Fichtner, W., & Rentz, O. (2003b). Financing Renewable Energy Projects Via Closed-End Funds: A German Case Study. *Renewable Energy, 28*(13), 2023–2036. https://doi.org/10.1016/S0960-1481(03)00080-6

Fischer, B., & Wetzel, H. (2018). *Entwicklungsdynamik und strukturelle Merkmale von Energiegenossenschaften in Deutschland in den Jahren von 2009 bis 2015* (klimaGEN Working Paper No. 1). Kassel. Retrieved from Universität Kassel, Fachgebiet Volkswirtschaftslehre mit Schwerpunkt dezentrale Energiewirtschaft website: https://www.uni-kassel.de/fb07/fileadmin/datas/fb07/5-Institute/IVWL/Wetzel/KlimaGEN_Arbeitspapier_1_web.pdf

Fleiß, E., Hatzl, S., Seebauer, S., & Posch, A. (2017). Money, Not Morale: The Impact of Desires and Beliefs on Private Investment in Photovoltaic Citizen Participation Initiatives. *Journal of Cleaner Production, 141*, 920–927. https://doi.org/10.1016/j.jclepro.2016.09.123

Fraunhofer Institute for Solar Energy Systems. (2021, January 4). *Public Net Electricity Generation in Germany 2020: Share from Renewables Exceeds 50 percent* [Press release]. Freiburg. Retrieved from https://www.ise.fraunhofer.de/en/press-media/news/2020/public-net-electricity-generation-in-germany-2020-share-from-renewables-exceeds-50-percent.html

Grashof, K. (2019). Are Auctions Likely to Deter Community Wind Projects? And Would This be Problematic? *Energy Policy, 125*, 20–32. https://doi.org/10.1016/j.enpol.2018.10.010

Grashof, K., Guss, H., Weiler, K., Matschoss, P., Zweck, B., Holstenkamp, L., & Möller, J. (2019). *Detailauswertung der ersten Ausschreibungsrunde für Windenergie an Land: Implikationen für die Entwicklung des Monitorings von Akteursvielfalt* (Climate Change No. 32/2019). Dessau. Retrieved from https://www.umweltbundesamt.de/sites/default/files/medien/1410/publikationen/2019-08-30_climate-change_32-2019_akteursvielfalt-photovoltaic-windenergie__0.pdf

Gugerli, D. (1996). *Redeströme: Zur Elektrifizierung der Schweiz; 1880–1914*. Zugl.: Zürich, Univ., Habil.-Schr., 1994/95. Chronos.

Hager, C., & Hamagami, N. (2020). Local Renewable Energy Initiatives in Germany and Japan in a Changing National Policy Environment. *Review of Policy Research, 37*(3), 386–411. https://doi.org/10.1111/ropr.12372

Hall, S., Roelich, K. E., Davis, M. E., & Holstenkamp, L. (2018). Finance and Justice in Low-Carbon Energy Transitions. *Applied Energy, 222*, 772–780. https://doi.org/10.1016/j.apenergy.2018.04.007

Hein, W., & Holstenkamp, L. (2018). Global Governance and the Interplay of Coordination and Contestation: The Case of Renewable Energies in the South. *Global Governance: A Review of Multilateralism and International Organizations, 24*(3), 431–449. https://doi.org/10.1163/19426720-02403008

Herbes, C., Brummer, V., Rognli, J., Blazejewski, S., & Gericke, N. (2017). Responding to Policy Change: New Business Models for Renewable Energy Cooperatives—Barriers Perceived by Cooperatives' Members. *Energy Policy, 109*, 82–95. https://doi.org/10.1016/j.enpol.2017.06.051

Holstenkamp, L. (2014). Local Investment Schemes for Renewable Energy: A Financial Perspective. In M. Peeters & T. Schomerus (Eds.), Renewable Energy Law in the EU: Legal Perspectives on Bottom-Up Approaches (pp. 232–255). Edward Elgar Publishing; Elgar. https://doi.org/10.4337/9781783473199.00021.

Holstenkamp, L. (2015a). Fallstudie 1: Die Inhaberschuldverschreibung von Hamburg Energie Solar. In C. Herbes & C. Friege (Eds.), *Handbuch Finanzierung von Erneuerbare-Energie-Projekten* (pp. 403–414). UVK, Konstanz & Munich.

Holstenkamp, L. (2015b). *The Rise and Fall of Electricity Distribution Cooperatives in Germany. Arbeitspapierreihe Wirtschaft & Recht: Vol. 22.* Leuphana Universität Lüneburg, Institut für Bank- Finanz- und Rechnungswesen (IBFR). Retrieved from https://www.leuphana.de/fileadmin/user_upload/Forschungseinrichtungen/ifa/professuren/finanzierung-finanzwirtschaft/files/Arbeitspapiere/wpbl22.pdf

Holstenkamp, L. (2018). Die Geschichte der Elektrizitätsgenossenschaften in Deutschland. In L. Holstenkamp & J. Radtke (Eds.), *Handbuch Energiewende und Partizipation* (pp. 405–421). Springer VS.

Holstenkamp, L., Centgraf, S., Dorniok, D., Kahla, F., Masson, T., Müller, J. R., & Yildiz, Ö. (2018). Bürgerenergiegesellschaften in Deutschland. In L. Holstenkamp & J. Radtke (Eds.), *Handbuch Energiewende und Partizipation* (pp. 1057–1075). Springer VS.

Holstenkamp, L., & Degenhart, H. (2014). Problemfelder und mögliche Lösungsansätze bei genossenschaftlichen Bürgerwindparks: Ressourcenmobilisierung und Projektaquise. *Zeitschrift für das gesamte Genossenschaftswesen, 64*(3), 185–200. https://doi.org/10.1515/zfgg-2014-0304

Holstenkamp, L., & Kahla, F. (2016). What Are Community Energy Companies Trying to Accomplish?: An Empirical Investigation of Investment Motives in the German Case. *Energy Policy, 97*, 112–122. https://doi.org/10.1016/j.enpol.2016.07.010

Holstenkamp, L., & Stier, B. (2018). Dezentrale Energieversorgung in Deutschland in historischer Perspektive. In L. Holstenkamp & J. Radtke (Eds.), *Handbuch Energiewende und Partizipation* (pp. 387–401). Springer VS.

Horstink, L., Wittmayer, J. M., Ng, K., Luz, G. P., Marín-González, E., Gährs, S., & Brown, D. (2020). Collective Renewable Energy Prosumers and the Promises of the Energy Union: Taking Stock. *Energies, 13*(2), 421. https://doi.org/10.3390/en13020421

Hughes, T. P. (1983). *Networks of Power: Electrification in Western Society, 1880–1930*. Johns Hopkins University Press.

Jacobsson, S., & Lauber, V. (2006). The Politics and Policy of Energy System Transformation: Explaining the German Diffusion of Renewable Energy Technology. *Energy Policy, 34*(3), 256–276. https://doi.org/10.1016/j.enpol.2004.08.029

Jenssen, T., König, A., & Eltrop, L. (2014). Bioenergy Villages in Germany: Bringing a Low Carbon Energy Supply for Rural Areas into Practice. *Renewable Energy, 61*, 74–80. https://doi.org/10.1016/j.renene.2012.08.014

Jordheim, H. (2012). Against Periodization: Koselleck's Theory of Multiple Temporalities. *History and Theory, 51*(2), 151–171. https://doi.org/10.1111/j.1468-2303.2012.00619.x

Kahla, F., Holstenkamp, L., Müller, J. R., & Degenhart, H. (2017). *Entwicklung und Stand von Bürgerenergiegesellschaften und Energiegenossenschaften in Deutschland* (Arbeitspapierreihe Wirtschaft & Recht No. 27). Lüneburg. Retrieved from Leuphana Universität Lüneburg (Leuphana), Institut für

Bank , Finanz- und Rechnungswesen website: http://www.leuphana.de/fileadmin/user_upload/Forschungseinrichtungen/professuren/finanzierungfinanzwirtschaft/files/Arbeitspapiere/wpbl27_BEG-Stand_Entwicklungen.pdf
Karapin, R. (2020). Household Costs and Resistance to Germany's Energy Transition. *Review of Policy Research, 37*(3), 313–341. https://doi.org/10.1111/ropr.12371
Klagge, B., & Meister, T. (2018). Energy cooperatives in Germany—An example of successful alternative economies? *Local Environment, 23*(7), 697–716. https://doi.org/10.1080/13549839.2018.1436045
Krause, F., Bossel, H., & Mueller-Reissmann, K. F. (1980). *Energie-Wende: Wachstum u. Wohlstand ohne Erdöl u. Uran; e. Alternativ-Bericht d. Öko-Inst., Freiburg*. S. Fischer.
Krauss, W. (2010). The 'Dingpolitik' of Wind Energy in Northern German Landscapes: An Ethnographic Case Study. *Landscape Research, 35*(2), 195–208. https://doi.org/10.1080/01426390903557972
Leiren, M. D., & Reimer, I. (2018). Historical Institutionalist Perspective on the Shift from Feed-In Tariffs Towards Auctioning in German Renewable Energy Policy. *Energy Research & Social Science, 43*, 33–40. https://doi.org/10.1016/j.erss.2018.05.022
Lieberman, E. S. (2001). Causal Inference in Historical Institutional Analysis: A Specification of Periodization Strategies. *Comparative Political Studies, 34*(9), 1011–1035. https://doi.org/10.1177/2F001041400103400900 3
Lienhoop, N. (2018). Acceptance of Wind Energy and the Role of Financial and Procedural Participation: An Investigation with Focus Groups and Choice Experiments. *Energy Policy, 118*, 97–105. https://doi.org/10.1016/j.enpol.2018.03.063
Lorenz, C. (2017). 'The Times They Are a-Changin': On Time, Space and Periodization in History. In M. Carretero, S. Berger, & M. Grever (Eds.), *Palgrave handbooks. Palgrave Handbook of Research in Historical Culture and Education* (pp. 109–131). Palgrave Macmillan.
Lovins, A. B. (1977). *Soft Energy Paths: Toward a Durable Peace. Pelican Books*. Penguin Books.
Maly, C. (2014). Legal Aspects of Local Engagement: Land Planning and Citizens' Financial Participation in Wind Energy Projects. In M. Peeters & T. Schomerus (Eds.), *Renewable Energy Law in the EU: Legal Perspectives on Bottom-Up Approaches* (pp. 210–231). Elgar.
Markowitz, L., & Tice, K. W. (2002). Paradoxes of Professionalization. *Gender & Society, 16*(6), 941–958. https://doi.org/10.1177/089124302237896

Mautz, R., Byzio, A., & Rosenbaum, W. (2008). *Auf dem Weg zur Energiewende: Die Entwicklung der Stromproduktion aus erneuerbaren Energien in Deutschland: eine Studie aus dem Soziologischen Forschungsinstitut Göttingen (SOFI)*. Universitätsverlag Göttingen. Retrieved from http://www.oapen.org/search?identifier=610261

Mori, P. A. (2013). Customer Ownership of Public Utilities: New Wine in Old Bottles. *Journal of Entrepreneurial and Organizational Diversity, 2*(1), 54–74. https://doi.org/10.5947/jeod.2013.004

Morris, C., & Jungjohann, A. (2016). *Energy Democracy: Germany's Energiewende to Renewables*. Springer International Publishing. Retrieved from https://doi.org/10.1007/978-3-319-31891-2.

Nolden, C., Barnes, J., & Nicholls, J. (2020). Community energy business model evolution: A review of solar photovoltaic developments in England. *Renewable and Sustainable Energy Reviews, 122*, 109722. https://doi.org/10.1016/j.rser.2020.109722

Olsen, B. E. (2014). Regulatory Financial Obligations for Promoting Local Acceptance of Renewable Energy Projects. In M. Peeters & T. Schomerus (Eds.), *Renewable Energy Law in the EU: Legal Perspectives on Bottom-Up Approaches* (pp. 189–209). Edward Elgar Publishing; Elgar.

Pellegrini, L., & Tasciotti, L. (2012). Rural Electrification Now and Then: Comparing Contemporary Challenges in Developing Countries to the USA's Experience in Retrospect. *Forum for Development Studies, 40*(1), 153–176. https://doi.org/10.1111/area.12444

Roesler, T. (2019). Community Resources for Energy Transition: Implementing Bioenergy Villages in Germany. *Area, 51*(2), 268–276. https://doi.org/10.1007/978-1-137-26317-9_40

Rothschild, J., Chen, K. K., Smith, D. H., & [David H.], & Kristmundsson, O. (2016). Avoiding Bureaucratization and Mission Drift in Associations. In D. H. Smith, R. A. Stebbins, & J. Grotz (Eds.), *The Palgrave Handbook of Volunteering, Civic Participation, and Nonprofit Associations* (pp. 1007–1024). Palgrave Macmillan. https://doi.org/10.1007/978-1-137-26317-9_40

Rübsamen, R., Delfs, C., Haas, G., & Lassen, R. (1995). *Energiegemeinschaften: Umweltfreundliche Stromerzeugung in der Praxis*. Piper.

Sabatier, P. A. (1987). Knowledge, Policy-Oriented Learning, and Policy Change: An Advocacy Coalition Framework. *Knowledge, 8*(4), 649–692. https://doi.org/10.1177/2F0164025987008004005

Schmid, B., Meister, T., Klagge, B., & Seidl, I. (2020). Energy Cooperatives and Municipalities in Local Energy Governance Arrangements in Switzerland and Germany. *The Journal of Environment & Development, 29*(1), 123–146. https://doi.org/10.1177/2F1070496519886013

Schreuer, A., & Weismeier-Sammer, D. (2010). *Energy Cooperatives and Local Ownership in the Field of Renewable Energy Technologies: A Literature Review* (RiCC – Research report No. 2010/4). Retrieved from Research Institute for Co-operation and Co-operatives (RiCC) website: https://epub.wu.ac.at/2897/1/Literature_Overview_energy_cooperatives_final_(2).pdf

Stefes, C. H. (2016). Critical Junctures and the German Energiewende. In C. Hager & C. H. Stefes (Eds.), *Germany's Energy Transition* (pp. 63–89). Palgrave Macmillan US. https://doi.org/10.1057/978-1-137-44288-8_3

Stefes, C. H. (2020). Opposing Energy Transitions: Modeling the Contested Nature of Energy Transitions in the Electricity Sector. *Review of Policy Research, 37*(3), 292–312. https://doi.org/10.1111/ropr.12381

Stier, B. (1999). *Staat und Strom: Die politische Steuerung des Elektrizitätssystems in Deutschland 1890–1950*. Univ., Habil.-Schr.—Mannheim, 1997. *Technik + Arbeit: Vol. 10*. Verl. Regionalkultur.

Süsser, D., Döring, M., & Ratter, B. M. W. (2017). Harvesting Energy: Place and Local Entrepreneurship in Community-Based Renewable Energy Transition. *Energy Policy, 101*, 332–341. https://doi.org/10.1016/j.enpol.2016.10.018

Süsser, D., & Kannen, A. (2017). 'Renewables? Yes, Please!': Perceptions and Assessment of Community Transition Induced by Renewable-Energy Projects in North Frisia. *Sustainability Science, 12*(4), 563–578. https://doi.org/10.1007/s11625-017-0433-5

Tews, K. (2018). The Crash of a Policy Pilot to Legally Define Community Energy. Evidence from the German Auction Scheme. *Sustainability, 10*(10), 3397. https://doi.org/10.3390/su1010339

Toke, D. (2011). *Ecological Modernisation and Renewable Energy. Energy, Climate, and the Environment Series*. Palgrave Macmillan.

Walker, G., & Devine-Wright, P. (2008). Community Renewable Energy: What Should it Mean? *Energy Policy, 36*(2), 497–500. https://doi.org/10.1016/j.enpol.2007.10.019

Weiler, K., Weber, A., Grashof, K., Holstenkamp, L., & Ehrtmann, M. (2021). *Entwicklung und Umsetzung eines Monitoringsystems zur Analyse der Akteursstruktur bei Freiflächen-Photovoltaik und der Windenergie an Land: Abschlussbericht: Ergebnisse des Monitorings und Empfehlungen* (Climate Change). German Environment Agency, Dessau. Retrieved from https://www.umweltbundesamt.de/sites/default/files/medien/5750/publikationen/2021-06-28_cc_49-2021_monitoringsystem_akteursstruktur_wind_pv.pdf

Weiler, K., Weber, A., Grashof, K., Matschoss, P., Klann, U., Hildebrand, J., & Möller, J. (2019a). *Entwicklung und Umsetzung eines Monitoringsystems zur Analyse der Akteursstruktur bei Freiflächen-Photovoltaik und der Windenergie an Land: Teilbericht: Methodik zur Erhebung der Akteursstruktur* (Climate Change). Dessau. Retrieved from https://www.umweltbundesamt.de/themen/klima-energie/erneuerbare-energien/erneuerbare-energien-gesetz/akteursstruktur-beim-ausbau-der-erneuerbaren

Weiler, K., Weber, A., Holstenkamp, L., & Ehrtmann, M. (2019b). *Kurzbericht zur Akteursstruktur der Ausschreibungen für Windenergie an Land*. Mai 2017–Mai 2019. Dessau. Retrieved from https://www.umweltbundesamt.de/themen/klima-energie/erneuerbare-energien/erneuerbare-energien-gesetz/akteursstruktur-beim-ausbau-der-erneuerbaren

Weiler, K., Weber, A., Holstenkamp, L., & Ehrtmann, M. (2020a). *Kurzbericht zur Akteursstruktur der Ausschreibung für Photovoltaik-Anlagen*. Februar 2017–Juni 2019. Dessau. Retrieved from https://www.umweltbundesamt.de/themen/klima-energie/erneuerbare-energien/erneuerbare-energien-gesetz/akteursstruktur-beim-ausbau-der-erneuerbaren

Weiler, K., Weber, A., Holstenkamp, L., & Ehrtmann, M. (2020b). *Kurzbericht zur Akteursstruktur der Ausschreibung für Photovoltaik-Freiflächenanlagen*. April 2015–August 2016. Dessau. Retrieved from https://www.umweltbundesamt.de/themen/klima-energie/erneuerbare-energien/erneuerbare-energien-gesetz/akteursstruktur-beim-ausbau-der-erneuerbaren

Wolman, P. (2007). The New Deal for Electricity in the United States, 1930–1950. In D. F. Barnes (Ed.), *The Challenge of Rural Electrification: Strategies for Developing Countries* (pp. 259–292). RFF Press.

Wüste, A., & Schmuck, P. (2012). Bioenergy Villages and Regions in Germany: An Interview Study with Initiators of Communal Bioenergy Projects on the Success Factors for Restructuring the Energy Supply of the Community. *Sustainability, 4*(2), 244–256. https://doi.org/10.3390/su4020244

Yildiz, Ö., Gotchev, B., Holstenkamp, L., Müller, J. R., Radtke, J., & Welle, L. (2019). Consumer (Co-)ownership in Renewables in Germany. In J. Lowitzsch (Ed.), *Energy Transition: Financing Consumer Co-Ownership in Renewables* (pp. 271–293). Springer/Palgrave Macmillan. https://doi.org/10.1007/978-3-319-93518-8_13

Yildiz, Ö., Rommel, J., Debor, S., Holstenkamp, L., Mey, F., Müller, J. R., & Rognli, J. (2015). Renewable Energy Cooperatives as Gatekeepers or Facilitators?: Recent Developments in Germany and a Multidisciplinary Research Agenda. *Energy Research & Social Science, 6*, 59–73. https://doi.org/10.1016/j.erss.2014.12.001

7

Support Structures for Renewable Energy Communities

Beau Warbroek

7.1 Introduction

The new Clean Energy Directive and the transposition of the new directives by the Member States does not only aim at legislation to incorporate the new role of citizens and communities in their energy legislation. It also aims to strengthen Member States' policies to support the empowerment of smaller and non-commercial market actors in the energy market. Policies to support energy communities were already in place in some Member States on different government levels. However, particularly local governments have problems to cope with the dynamics of the development and upscaling of renewable energy communities (RECs) due to their risk-adverse culture and inadequate resources for administrative capacity have problems to cope with the dynamics of the development and upscaling of renewable energy communities (RECs) (Voorberg et al., 2014).

B. Warbroek (✉)
Department of Civil Engineering, University of Twente,
Enschede, The Netherlands
e-mail: w.d.b.warbroek@utwente.nl

© The Author(s), under exclusive license to Springer Nature Switzerland AG 2021

F. H. J. M. Coenen, T. Hoppe (eds.), *Renewable Energy Communities and the Low Carbon Energy Transition in Europe*, https://doi.org/10.1007/978-3-030-84440-0_7

To further REC development, support is needed to build their capacities; institutional hurdles and barriers stemming from the fossil fuel-based energy regime need to be alleviated ; and the energy systems needs to open up for the uptake, acceptance or breakthrough of RECs. Evidence suggests that so-called "intermediaries" form a part of the solution in addressing these issues. Although intermediaries can be different kinds of actors including governments themselves, I will argue that non-governmental actors are most effective in performing the intermediary tasks due to the nature of this work. This work asks intermediaries to bridge dividing lines and try to overcome institutional inertia to which governments are often prone.

On the basis of the question of this volume: "How did renewable energy communities in continental Europe evolve, and what we can learn from these experiences for the realization and implementation of the Clean Energy Package (CEP)?", this chapter focuses on the experiences with support of RECs with an emphasis on the support by so-called intermediaries. This chapter presents an analytical framework on the various roles and strategies intermediaries can employ to support RECs. It addresses the question: What strategies, roles and activities of intermediaries represent an effective support structure for the further development and upscaling of RECs? This chapter builds on an extensive literature review and experiences from the case of the Dutch province of Fryslân (Warbroek et al., 2018; Warbroek, 2019).

7.2 Development of Dutch Renewable Energy Communities

Motivated by the energy crises of the 1970s, the report of the Club of Rome ("Limits to growth"), the rise of the anti-nuclear energy movement, and the attention that resulted from this to generate energy from sustainable, clean sources, the first citizen energy collectives were created in the Netherlands with mainly wind energy projects, such as the Windvogel. Between 1995 and 2009 there were no more than twenty energy collectives active in the Netherlands (Oteman et al., 2014;

Schwencke, 2017). Only after 2010 did their number increase significantly. Partly stimulated by falling prices for generating solar power and a favourable incentive scheme for generating renewable energy (SDE+), and the advent of organizational and capacity support measures from local and provincial governments saw a growing number of energy collectives come into being.

Although several energy production technologies were central to business models of these cooperatives—solar, wind, biogas, heat, energy saving—they mostly focused on solar energy, with a lot of attention being paid to purchasing for households and developing collective solar farms. From 2013 (from the national Energy Agreement) the development of solar power was supported by the central government through the so-called 'postcode rose' scheme, a tax measure which allows cooperatives (and associations of home owners) to generate sustainable energy on a small scale with a common installation (and are exempt from VAT and energy tax) (Kooij et al., 2018). As discussed in the introduction chapter the number of these cooperatives has grown quickly in the last years, moving the Netherlands from the middle group of countries to a front-runner in 2020 with 623 initiatives (Schwencke, 2021).

Nevertheless, it can be said that the growth in the number of energy collectives has flattened in recent years (Bauwens et al., 2016; Schwencke, 2021). Reasons given for this in the Netherlands are mainly related to possibilities for installation of solar PV, such as the decreasing availability of roofs, problems with grid capacity, phasing out of the netting scheme, uncertainty around adjustment of support measures (both fiscally and in terms of subsidy) and the corona virus pandemic (which makes it more difficult for energy cooperatives to recruit members) (ibid.). A similar development is taking place in other countries, such as Germany (Bauwens et al., 2016).

Most support instruments for the integration of renewables are designed at the national level, although the liberalization of the energy sector in Europe has ensured that energy cooperatives could also engage in for example the supply of energy to their members. This did not lead to an equal energy market position for RECs. As discussed in Chap. 2 of this volume, the CEP creates the building blocks for a fair and equal treatment in national legislation. The definition of REC (and Citizen

Energy Community, CEC) not only recognizes the position of these communities as a legal entity that organizes itself around specific ownership, democratic governance and a non-commercial objective, but also ensures that citizens' initiatives are not excluded from the market. For instance, Member States get the opportunity to make different arrangements for energy communities and small-scale installations outside the standard tender procedures (such as the SDE+ in the Netherlands). The European legislation needs to be transposed into Dutch law. At the moment of writing, it was not yet clear exactly how this will take shape and what the significance is for the Dutch energy cooperatives. However, it is already clear that the Dutch energy cooperatives generally meet this European definition(s). In 2020, the Dutch cooperatives introduced the concept of a heat-energy community (*warmteschap* in Dutch) which is closely aligned with the European definition. In the Netherlands on 17 December 2020 the responsible Ministry of Economic Affairs submitted the Energy Act bill for consultation. This Energy Act bill must implement the European regulations regarding the internal energy market and the Clean Energy Package and implement Dutch agreements from the so-called Climate Agreement and other national policy intentions. The current Electricity Act from 1998 from contained experimental space for (smart) grid operators and RECs as a form of regulatory 'sandbox'.

Many factors that influence the survival and growth of RECs are not only related to their position in the energy market but also to their specific internal characteristics. This chapter focuses mostly upon the individual characteristics of RECs, their organization, their relation with the local community, their supportive environment and relation with the existing actors in the energy market in the period before the conclusion of the CEP. All these characteristics are related with factors that influence the survival of citizen initiatives in the energy transition. These factors lead to particular barriers in terms of (resource) shortcomings of RECs and unsupportive institutional settings for the development and upscaling of RECs.

7.3 Factors Related to the Survival of RECs in the Energy Transition

The academic literature shows that the survival of citizen energy collectives is (positively) related to a number of factors. In short, it concerns the (i) organization of the energy collective, (ii) the interaction with the local environment and (iii) the extent to which the collective is supported by the policy environment (Warbroek et al., 2018). As a fourth category of influence, the interaction and dependence of the incumbent 'regime' can also be discerned.

With regard to the organization of the REC, the following factors matter for success: leadership, the presence of different types of human capital (Ornetzeder & Rohracher, 2013), size (the core group of active volunteers) (Feola & Nunes, 2014), diversity (not only a homogeneous composition with highly educated, native middle- and older-age men) (Van Der Schoor & Scholtens, 2015), time that members can make available (Feola & Nunes, 2014) and financial capacity (Ornetzeder & Rohracher, 2013). It is important to stress is that the organization usually relies on volunteering and the commitment of a small core group that coordinates the ins and outs of the collective. Professionalization of and dependence on members of this core group are sensitivities that many collectives face. When the core group disintegrates, this can cause the whole collective to fall over (unless timely follow-up is found).

Within the cluster interaction with the local community, factors such as local values, structure and rules of the community (Süsser et al., 2017) are considered important. Involvement and visibility within that community also play a role (Boon & Dieperink, 2014). This touches upon the importance of social capital as well. This includes, on the one hand, attachment, sharing social norms with, and enjoying trust among the local community (having so-called 'strong bonds') (Walker et al., 2010). On the other hand, there are also different, less intense and frequent links with local and other communities and organizations (the so-called weaker links) (Granovetter, 1973). In practice, having both strong and weak social connections is important for energy collectives (Martiskainen, 2017).

Third, a supportive policy context is important. This concerns the relationship to, and support by, governments and other intermediary parties (Ruggiero et al., 2014). Examples of such support include a municipality that makes a venue available so that the energy collective can organize meetings, or that helps the collective to apply for a grant to generate sustainable energy. These may also be regional foundations that help energy collectives in formation to draw up the formal statutes for association or to develop a business model (Warbroek et al., 2018). For the feasibility of the business model of energy collectives, sufficient policy support is key. This includes, for example, (the aforementioned) fiscal or subsidization measures that make investments in renewable energy installations more profitable (Kooij et al., 2018; Oteman et al., 2014).

Finally, it is also important to uncover how traditional parties in energy markets interact with RECs. Although energy collectives cannot compete with energy companies, grid operators and governments in their projects, the interests of these parties are not always the same as those of the citizens' initiatives. A study by Proka et al. (2018) shows that the independence of energy collectives is limited by the behaviour of these ('regime') parties. This is mainly due to the way in which they limit the business models of energy collectives by shaping conditions around supportive policy measures in such a way that they are more favourable to themselves than to civilian collectives. This has been done through the exercise of political power, institutional culture and the dominant knowledge system. For example, it appears that in the past the main subsidy scheme for renewable energy has mainly benefited traditional energy market parties and not citizens' initiatives, although the latter have also successfully applied for grants. In practice, this means that large-scale energy generation projects are preferred over decentralized, small-scale citizen-initiated energy projects.

Given these barriers, the further development of RECs largely depends on three focus areas: (i) capacity building and embedding of RECs; (ii) removing barriers and lock-ins; and (iii) opening up the existing institutional arrangements for the incorporation, acceptance or breakthrough of RECs. Given the nature of these challenges, the relevance of intermediary work becomes apparent.

7.4 Intermediaries as the Solution for the Problems of RECs

Indeed, several studies suggest that intermediaries might be part of the solution in addressing the above mentioned resource shortcomings and unsupportive institutional settings for the development and upscaling of RECs (e.g., Hargreaves et al., 2013; Seyfang et al., 2014). Intermediaries open up the existing system and introduce dynamism by creating opportunities and spaces for others (Howells, 2006, p. 726; Stewart & Hyysalo, 2008, pp. 296–297). Within these spaces and dynamics, intermediaries fulfil a network managing role, where they "mediate, work in between, make connections, and enable a relationship between different persons or things" (Hodson et al., 2013, p. 1408; Moss, 2009).

Studies on intermediaries reveal a wide variety of actors who can carry out intermediary activities, such as government agencies, NGOs, energy service companies (ESCOs), academic institutions, consultancies, councils, business networking platforms and individuals (Backhaus, 2010; Bird & Barnes, 2014; Kivimaa, 2014; Kivimaa & Martiskainen, 2017; Kivimaa & Martiskainen, 2018). Governments can also carry out intermediary activities in the form of a supportive mode of governance (Bulkeley & Kern, 2006; Hoppe et al., 2015; Warbroek & Hoppe, 2017). Intermediaries are therefore best conceptualized in terms of their activities and the processes they undertake, rather than who or what kind of actors perform these actions (Kivimaa & Martiskainen, 2018). As a consequence, various adjectives are used in the literature to categorize the functions of intermediaries, including transition intermediaries (Kivimaa et al., 2017), innovation intermediaries (Howells, 2006; Martiskainen & Kivimaa, 2018; Stewart & Hyysalo, 2008), energy intermediaries (Backhaus, 2010; Hodson et al., 2013), user intermediaries (Barnes, 2017; Boon et al., 2011), and niche intermediaries (Hargreaves et al., 2013; Seyfang et al., 2014).

Studies also show a wide variety in the roles that intermediaries can play and the different activities they can perform in relation to RECs (Bomberg & McEwen, 2012; Forrest & Wiek, 2014; Hargreaves et al., 2013; Hicks & Ison, 2011; Parag et al., 2013; Ruggiero et al., 2014;

Seyfang et al., 2014). Studies from different countries show that intermediaries play a key role in promoting knowledge transfer, information provision and capacity building; and are central to establishing partnerships between RECs and actors from outside the community energy sector, such as the incumbent operators of the regime and central actors of the energy system (Hargreaves et al., 2013; Parag et al., 2013; Ruggiero et al., 2014). Transition study authors have argued that intermediaries play a key role in the development of strategic niches (Geels & Raven, 2006).

There have been previous attempts to analyse intermediary roles and activities related to community energy development (e.g., Bird & Barnes, 2014; Hargreaves et al., 2013; Seyfang et al., 2014; Smith et al., 2015). The reality of the different roles and strategies that intermediaries can use and the support that RECs need for their development (i.e., capacity building and embedding RECs in their communities, removing barriers and opening up existing institutional arrangements) were summarized in a comprehensive analytical framework by Warbroek et al. (2019). The goal of the framework is to understand in what ways different strategies shape the roles and activities of intermediaries in supporting RECs. Although the primary purpose of intermediary support is to address the problems of the RECs discussed before, there are several ways intermediaries can do this.

7.5 Strategies that Intermediaries Use

The roles and activities of intermediaries are driven by strategies. It is argued here that the support of RECs should involve a combination of multiple strategies to successfully address the issues hindering RECs and promote their development. One of the most dominant perspectives underpinning such a strategy is that of Strategic Niche Management (SNM). SNM arose from research into socio-technical transitions. In the low-carbon energy transition, advocates argue for the need for social and technological innovations to cope with climate change (Geels et al., 2008). Radical innovations that potentially destabilize the existing socio-technical regime must be nurtured in protected spaces before they can spread further (Kemp et al., 1998). These protective spaces, also called niches, are protected against pressures of the existing regime (Smith and Raven 2012). SNM sheds light on the processes and strategies of how to create and develop niches to drive a system-wide transition (Kemp et al., 1998; Schot & Geels, 2008). The theory suggests that successful niches can influence the regime by enabling the replication of projects within the niche and driving change through multiple small initiatives; by growing sub-projects in size and attracting more participants; and by facilitating the translation of niche ideas into mainstream settings (Seyfang & Haxeltine, 2012). In the process of niche development, intermediaries seem to be of particular importance (Geels & Deuten, 2006; Hargreaves et al., 2013; Ruggiero et al., 2018; Ruggiero et al., 2014; Seyfang et al., 2014). Therefore, SNM mainly deals with issues related to the removal of barriers and opening up existing institutional arrangements to facilitate the acceptance and breakthrough of RECs.

The Business Incubator approach aims to accelerate the creation of successful entrepreneurial companies individually (Bruneel et al., 2012), without emphasizing the development of a specific niche. Instead, the incubator itself forms a protective space by providing start-ups with shared office spaces and equipment; administrative services (e.g., reception and administrative services); business support with coaching and training activities (i.e., business planning, marketing, accounting, management support and access to finance); and access to services through external networks (Bergek & Norrman, 2008; Bøllingtoft & Ulhøi, 2005; Bruneel et al., 2012; Hansen et al., 2000; Lai & Lin, 2015; Pauwels et al., 2016; Smilor, 1987). Business incubators help new ventures get past the early critical years (Bøllingtoft & Ulhøi, 2005) and accelerate their growth with the aim of creating self-sustaining, thriving businesses. In terms of REC support, the incubator approach focuses primarily on building the capacity of RECs for start-ups and removing barriers associated with that start-up phase.

Two other perspectives so far not associated with the community energy and intermediate literature also provide helpful suggestions for intermediary strategies. The findings of various studies showing that existing or potential internal capabilities and (symbolic) resources are crucial in community-based bottom-up developments (Bomberg & McEwen, 2012; Forrest & Wiek, 2014; Middlemiss & Parrish, 2010;

Walker & McCarthy, 2010) and can be traced back to the ideas of Endogenous Development and Asset-Based Community Development..

Initially introduced as an innovative approach to rural development, the key principle of Endogenous Development is that development will be more successful and sustainable if it: (i) draws on local resources; and (ii) includes public participation in the design and implementation phase (Ray, 1999b, p.524; see also Shucksmith, 2000). As such, endogenous development builds on, stimulates and supports social innovation (Neumeier, 2012, p. 59). The endogenous development approach connects people and their innovations, entrepreneurship and capital to place. Endogenous development is about "the idea of local ownership of resources and the sense of choice (local, collective agency) to use those resources (physical and immaterial) in pursuit of local goals" (Ray, 1999b, p. 525).

Within this, cultural-territorial identity is a central resource for RECs to draw upon. Ray (1999a) argues that culture (e.g., language and dialect, local knowledge, landscape) can be used to stimulate and define development. In this sense, Bomberg and McEwen (2012) in the context of RECs showed that the mobilization of the community culture, values and identity can sustain community mobilization. Likewise, Forrest and Wiek (2014) noted that a significant solidarity from a common village identity and a sense of pride were also critical success factors. Ray (1998) suggests that local initiatives can use these cultural and identity symbols to reassess place and locate economic control. As such, the endogenous development approach seeks to promote the development of RECs by ensuring (local) ownership, participation and embedding by attaching place and local economy to RECs (see for instance Devine-Wright, 2013).

The Asset-Based Community Development (ABCD) approach was developed in response to socio-economic problems in American cities in the 1990s. The core axiom of the ABCD approach is to focus on the assets and capabilities of the community, rather than on its needs, shortcomings and problems in the community revitalization efforts (Kretzmann & McKnight, 1996). As a result, ABCD leaves the control to the initiators themselves and builds trust in the communities (Mathie & Cunningham, 2003). In addition, ABCD assumes that the development process is relationship-driven, making use of the social capital present in the community (Kretzmann & McKnight, 1996; Mathie & Cunningham, 2003). As a strategy, ABCD supports RECs by boosting the existing capabilities and assets of the local communities in which RECs are located.

Table 7.1 provides an overview of the above strategies. Intermediate support structures using all of the strategies listed in Table 7.1 are likely to be more successful in supporting RECs and promoting their development. Several strategies can be distinguished within a particular

Table 7.1 Overview of intermediary strategies

Strategy	Theoretical rationale	Assumptions regarding needs RECs have	Type of support to RECs
Strategic niche management	Build and nurture a protective space for individual experimentation and technological innovations to increase their potential to bring about a transition through replication, scaling up or translation. The niche is further developed by managing expectations, creating social networks and stimulating learning processes.	Niches and the experiments in them are seen as improvements to the existing regime. If niches cannot open up and influence the regime, they cease to exist. Support is aimed at further developing the niche.	Removing barriers, opening up systems for new practices
Business incubator	New ventures are given the resources and capabilities to accelerate their growth and enhance their survival.	Support is needed in the start-up phase of the REC, after which it is expected to survive on its own. Help is supported toward development of business models.	Removing barriers, building practical capacities

(*continued*)

Table 7.1 (continued)

Strategy	Theoretical rationale	Assumptions regarding needs RECs have	Type of support to RECs
Endogenous development	Using local resources, stakeholders and markers stemming from the territorial-identity to revitalize the locality. People at the local level know best how to tackle local problems. Stimulate social innovation.	REC support is contextualized and ensues by means of popular participation and ownership in its design and implementation.	Building endogenous capacities and embedding REC in its social context
Asset-based community development (ABCD)	Development is based on the capacities and assets that are present, instead of a focus on needs and deficiencies.	RECs need to be supported by focusing on existing community assets.	Building and using existing assets and capacities

Based on Warbroek et al. (2019)

intermediary role or activity. The different strategies I have outlined help to highlight the key assumptions that guide the roles and activities of the intermediaries, and help assess the completeness of intermediaries' support structure. The next section will discuss the different roles and activities of intermediaries.

7.6 Roles and Activities of Intermediaries

As with the variety of actors that can act as intermediaries and the different strategies that can be employed, the roles and activities of intermediaries also vary greatly. Below the different roles of intermediaries in accordance with the support that RECs need for their development are discussed.

7.6.1 Building Capacity and Embedding RECs

In distinguishing intermediary roles within innovation processes, Stewart and Hyysalo (2008) argue that intermediaries can play a facilitating role in building capacities for other actors. This role includes the collection and distribution of financial, technical and institutional resources and the provision of skills and knowledge. Howells (2006) noted that intermediaries help find funding and process, generate and combine knowledge. This role is characterized by activities such as conducting feasibility studies (Ruggiero et al., 2014) or providing technical and legal advice, as well as guidance on funding sources and applications (Bird & Barnes, 2014; Bomberg & McEwen, 2012; Dinnie & Holstead, 2017; Hinshelwood, 2001; Kivimaa, 2014; Ruggiero et al., 2014). The facilitating role is further illustrated by endogenous development, asset-based capacity building and embedding measures through: updating participants' personal, organizational and entrepreneurial capacities to ensure the survival of project advocates, developing complementary business initiatives (enhancing the financial viability of RECs); connecting and developing relationships with key figures in the sector; reviving the sense of community; and ensuring ownership (Guerreiro & Botetzagias, 2018). In addition, this role also includes facilitating and organizing interactions between RECs). Researchers further found that intermediaries facilitate access to information, as well as information flows and interactions between RECs to share experiences (Bird & Barnes, 2014; Parag et al., 2013; Hicks & Ison, 2011; Wade et al., 2013). In addition, personal mentorship and training workshops seem to help the RECs in particular (see also Hicks & Ison, 2011; Seyfang et al., 2014).

Geels and Deuten (2006) argue that intermediaries aggregate, distribute and convert local knowledge into robust, abstracted and standardized knowledge that can be shared between local practices, by means of case studies, toolkits and handbooks (Hargreaves et al., 2013) or common grant application templates (Bird & Barnes, 2014). Matschoss and Heiskanen (2017, p. 89) state that intermediaries may aggregate experiences and knowledge by hosting co-creation events, attracting non-local, expert and scientific knowledge and by collecting knowledge and experiences from other countries or experiments.

7.6.2 Removing Barriers

Brokering activities are crucial to overcome barriers. Brokering activities indicate the network manager role of intermediaries in innovation processes (see also Howells, 2006; Kivimaa, 2014). Intermediaries bring relevant actors into the innovation network, maintain their commitment and interest, and guarantee a certain degree of openness of the innovation network to other interests. In addition, brokering activities include negotiating on behalf of individuals and institutions that adopt the innovation. Hargreaves et al. (2013) also envisioned a brokering role for intermediaries in the community energy sector. Within the role of partnership formation and management, intermediaries introduce community initiatives to potential partners, mediate partnerships between community energy groups and large corporations, and protect the interests of community energy groups in the process. In the same vein, Matschoss and Heiskanen (2017) argue that intermediaries can challenge established practices by introducing new actor configurations. Intermediaries also engage in lobbying activities to influence policy- and decision-making. These activities are usually undertaken in the light of obtaining new sources of investment (Anglin & Montezemolo, 2004) and developing new business proposals for community energy groups. Intermediaries also have a representative function to outsiders (see also Hasanov & Zuidema, 2018), as they engage with policy-makers to address how policies influence the reality of RECs (Parag et al., 2013) thereby linking RECs to policies (Bird & Barnes, 2014, p. 213), and provide a channel of communication between RECs and government (Bird & Barnes, 2014). Similarly, according to Wade et al. (2013) intermediaries bridge the gap between centralized formal structures (government, energy companies) and the decentralized nature of RECs.

7.7 Opening Up Institutional Arrangements for the Uptake, Acceptance or Breakthrough of RECs

A configuring role involves shaping the innovation by configuring the content of the innovation, such as setting rules for its use, prioritizing, fine-tuning and shaping particular applications, goals and form of projects, as well as the goals, expectations and needs of other stakeholders (Howells, 2006; Klerkx & Leeuwis, 2009; Stewart & Hyysalo, 2008). Guerreiro and Botetzagias (2018) conceptualized this as seeing that intermediaries can provide a social fit for a technology. Intermediary work that develops new financing and business models is also relevant here (Guerreiro & Botetzagias, 2018; Huijben & Verbong, 2013; Kivimaa & Martiskainen, 2018) as is the scaling-up of local initiatives to match the interest of funding agencies (Wade et al., 2013). In other words, configuring RECs along with the low-carbon energy applications and settings involved will generate public and market acceptance critical to their wider diffusion and development (i.e., Wolsink, 2012).

To further generate acceptance of RECs and increase the impact of RECs outside their local context, intermediaries can play a managing and coordinating role. Geels and Deuten (2006) argue that guidance, advice and templates underpin this role. In this role, intermediaries coordinate between actors in decision-making arenas to avoid lock-ins and make progress in terms of diffusion of innovative processes and activities. As an example, Bird and Barnes (2014 p. 213) noted that intermediaries helped develop a shared vision that transcended the day-to-day practice and activities of RECs and provided a systematic view of the community's energy sector. In addition, intermediaries can shape debates and discourses in different ways to achieve favourable results in decision-making processes (Hisschemöller & Sioziou, 2013). For example, Rohracher (2009) discovered that intermediaries are trying to restructure the energy markets by establishing green electricity labels. These labels are intended to provide guidance and transparency in the supply of green energy and to make visible the demand for such products (Rohracher, 2009, p. 2015).

Geels and Deuten (2006) pointed to the creation of an institutional infrastructure as another role for intermediaries that may facilitate the uptake of RECs. The creation of a shared institutional infrastructure facilitates the standardization and stabilization of an innovation to meet the demands and expectations of mainstream users (Geels & Deuten, 2006), and aims to establish the shared rules or development path for the community energy sector (Hargreaves et al., 2013). However, the authors of both studies interpret the institutional infrastructure primarily as a forum for storing, exchanging and disseminating knowledge. When understanding institutions as "the formal or informal procedures, routines, norms and conventions that are embedded in the organizational structure of the state system or political economy" (Hall & Taylor, 1996, p. 938), institutional infrastructure gets a different meaning. Here, the institutional infrastructure for community energy takes as a starting point concepts such as small-scale, distributed generation, local ownership, community benefits and decentralized decision-making. These concepts define the interaction between actors. Communities wishing to set up a REC can join this infrastructure to accelerate their development. Such an infrastructure that actively supports and legitimizes RECs' activities and objectives gives REC greater reach, enhanced capabilities and drives adoption.

An overview of the roles and activities of intermediaries is given in Table 7.2. The intermediary roles and accompanying activities listed in Table 7.2, together with the strategies summarized in Table 7.1, can best be seen as an ecosystem: completeness and consistency of the support structure positively influences RECs development and as such constitutes an effective support structure.

7.8 Intermediaries Within Existing and New Policy Arrangements

The previous sections discussed the strategies, roles and activities of intermediaries that represent an effective support structure for the further development and upscaling of RECs. What we can learn from these

Table 7.2 Overview of intermediary roles and activities

Relevant support required by RECs	Associated roles from literature	Activities
Building capacities and embedding into community	Facilitating	Distributing financial, technical, institutional, knowledge resources, providing advice, building capacity, and skills.
	Aggregation of knowledge	Developing toolkits, handbooks, and templates, and distributing these.
Alleviating barriers within the status quo	Brokering	Advocacy, negotiation with other parties, representative function, lobbying, engaging with policy-makers, introducing new actor configurations, and embedding in current policy frameworks. Identifying and challenging institutionalized practices.
Opening up the system for the uptake, acceptance or breakthrough of RECs	Creating institutional infrastructure	Setting up a supportive environment in which local initiatives are embedded and integrated, and which governs interactions and activities.
	Configuring	Embedding technology in the local community. Prioritizing or shaping certain uses of the technology, developing new (business) models, and engaging in pilots.
	Framing and coordinating	Articulating demand, framing discourses and debates, and coordinating between actors in decision-making processes.

Based on Warbroek et al. (2018); Warbroek (2019)

experiences with support of RECs and then particular support by so-called intermediaries for the realization and implementation of the Clean Energy Package?

The implementation of the CEP and its directives pave the way for a variety of national support measures to address the challenges RECs are facing. As discussed in Chap. 2 by Roberts the Renewable Energy Directive II (RED II) mentions the importance of removing barriers and building capacity (for governments to support RECs, but also access to funding and information).

Particularly in the context of RECs, intermediaries exist by virtue of existing (innovation) policy and institutional arrangements. Kivimaa and Martiskainen (2018) have shown that policy changes can halt the activities of intermediaries, giving governments a prominent role in enabling intermediary work. This makes intermediaries dependent on the political and financial resources made available to them. This means that when intermediaries emphasize their frontier role as a means of lobbying for unconventional practices that clashes with the status quo, they may lose access to networks of influence (i.e., government), which is an important resource for both themselves and their target audiences. Therefore, any institutional support for intermediaries which anchors their work, more than policy, is desirable.

While such a national level policy framework is an important supportive factor (Bomberg & McEwen, 2012; Oteman et al., 2014), a great deal of the supportive work actually takes place at the regional and local level, where RECs and intermediaries also predominantly operate. Even more so, the stop-and-go policy that long characterized the Dutch energy policy context (e.g., Agnolucci, 2007; Verbong et al., 2008; Dinica & Arentsen, 2001) is not a good predictor for the continued (national-level) support for intermediary work. The crucial local support structure, however, is heavily fragmented and susceptible to political and policy change (Warbroek & Hoppe, 2017; Warbroek et al., 2018). A case in point is the Dutch province of Fryslân, which has shown an effective support structure characterized by a coherent and comprehensive configuration of intermediaries' strategies, roles and activities (Warbroek et al., 2018). In 2020 however, the provincial grant enabling the intermediary support structure ended, leaving a vacuum and a desperate call from RECs for someone to step in. As such, the importance of intermediary work along with the continued need for RECs to be supported, demands a more solid basis than one that is left to fragmentation, governments' terms of office and policy instability.

While this lack of entrenchment can partially be alleviated with institutional bricolage (combining existing institutional elements for a new purpose) (Warbroek & Hoppe, 2017), the support for RECs, and thus the viability of intermediaries ought to follow a more programmatic, strategic approach (compare Proka et al., 2018). Both institutional bricolage and a programmatic approach for the support of RECs should be implemented by both local and regional authorities, as they both benefit from the work of intermediaries and both experience a lack of capacity (expertise, know-how) to effectively support RECs themselves.

While the above mentioned perspective leans heavily on the argument that government actors should step in to enable intermediary work, RECs in the province of Fryslân have shown that a supportive institutional structure can be created by the REC movement itself as well. By reselling to individual households green energy that is supplied by the regional energy supplier of which individual RECs are all shareholders, the RECs manage to keep the money that circulates with energy supply and demand in the region. This creates financial streams and provides a common 'face' to the REC movement. As such, the way the work of intermediaries can be entrenched should come from multiple dynamics (top-down, bottom-up) as well as multiple sectors (market, community, state).

References

Agnolucci, P. (2007). Renewable electricity policies in The Netherlands. *Renewable Energy, 32*(5), 868–883.

Anglin, R., & Montezemolo, S. (2004). Supporting the Community Development Movement: The Achievements and Challenges of Intermediary Organizations Roland. In R. Anglin (Ed.), *Building the Organizations That Build Communities* (pp. 55–27). Department of Housing and Urban Development.

Backhaus, J. (2010). Intermediaries as Innovating Actors in the Transition to a Sustainable Energy System. *Central European Journal of Public Policy, 4*(1), 86–108. Retrieved from http://www.cejpp.eu/index.php/ojs/article/view/47

Barnes, J. (2017). *User-Intermediaries and the Local Embedding of Low Carbon Technologies (2057-6668)* (Vol. 15) https://doi.org/2057-6668

Bauwens, T., Gotchev, B., & Holstenkamp, L. (2016). What Drives the Development of Community Energy in Europe? The Case of Wind Power Cooperatives. *Energy Research & Social Science, 13*, 136–147.

Bergek, A., & Norrman, C. (2008). Incubator Best Practice: A Framework. *Technovation, 28*(1–2), 20–28. https://doi.org/10.1016/j.technovation.2007.07.008

Bird, C., & Barnes, J. (2014). Scaling up Community Activism: The Role of Intermediaries in Collective Approaches to Community Energy. *People, Place and Policy Online, 8*(3), 208–221. https://doi.org/10.3351/ppp.0008.0003.0006

Bøllingtoft, A., & Ulhøi, J. P. (2005). The Networked Business Incubator – Leveraging Entrepreneurial Agency? *Journal of Business Venturing, 20*(2), 265–290. https://doi.org/10.1016/j.jbusvent.2003.12.005

Bomberg, E., & McEwen, N. (2012). Mobilizing Community Energy. *Energy Policy, 51*, 435–444. https://doi.org/10.1016/j.enpol.2012.08.045

Boon, F. P., & Dieperink, C. (2014). Local Civil Society Based Renewable Energy Organisations in the Netherlands: Exploring the Factors That Stimulate Their Emergence and Development. *Energy Policy, 69*, 297–307.

Boon, W. P. C., Moors, E. H. M., Kuhlmann, S., & Smits, R. E. H. M. (2011). Demand Articulation in Emerging Technologies: Intermediary User Organisations as Co-producers? *Research Policy, 40*(2), 242–252. https://doi.org/10.1016/j.respol.2010.09.006

Bruneel, J., Ratinho, T., Clarysse, B., & Groen, A. (2012). The Evolution of Business Incubators: Comparing Demand and Supply of Business Incubation Services Across Different Incubator Generations. *Technovation, 32*(2), 110–121. https://doi.org/10.1016/j.technovation.2011.11.003

Bulkeley, H., & Kern, K. (2006). Local Government and the Governing of Climate Change in Germany and the UK. *Urban Studies, 43*(12), 2237–2259. https://doi.org/10.1080/00420980600936491

Devine-Wright, P. (2013). Think Global, Act Local? The Relevance of Place Attachments and Place Identities in a Climate Changed World. *Global Environmental Change, 23*(1), 61–69.

Dinica, V., & Arentsen, M. J. (2001). *Green Electricity in the Netherlands*. Report no 2/2001, Norwegian School of Management, Centre for Energy and Environment.

Dinnie, E., & Holstead, K. L. (2017, January). The Influence of Public Funding on Community-Based Sustainability Projects in Scotland. *Environmental Innovation and Societal Transitions*, 1–9. https://doi.org/10.1016/j.eist.2017.08.003

Feola, G., & Nunes, R. (2014). Success and Failure of Grassroots Innovations for Addressing Climate Change: The Case of the Transition Movement. *Global Environmental Change, 24*, 232–250.

Forrest, N., & Wiek, A. (2014). Learning from success – Toward Evidence-informed Sustainability Transitions in Communities. *Environmental Innovation and Societal Transitions, 12*, 66–88. https://doi.org/10.1016/j.eist.2014.01.003

Geels, F., & Deuten, J. J. (2006). Local and Global Dynamics in Technological Development: A Sociocognitive Perspective on Knowledge Flows and Lessons from Reinforced Concrete. *Science and Public Policy, 33*(4), 265–275. https://doi.org/10.3152/147154306781778984

Geels, F., & Raven, R. (2006). Non-linearity and Expectations in Niche-Development Trajectories: Ups and Downs in Dutch Biogas Development (1973–2003). *Technology Analysis & Strategic Management, 18*(3–4), 375–392. https://doi.org/10.1080/09537320600777143

Geels, F. W., Hekkert, M. P., & Jacobsson, S. (2008). The Dynamics of Sustainable Innovation Journeys. *Technology Analysis & Strategic Management, 20*(5), 521–536. https://doi.org/10.1080/09537320802292982

Granovetter, M. S. (1973). The strength of weak ties. *American Journal of Sociology, 78*(6), 1360–1380.

Guerreiro, S., & Botetzagias, I. (2018). Empowering Communities–The Role of Intermediary Organisations in Community Renewable Energy Projects in Indonesia. *Local Environment, 23*(2), 158–177. https://doi.org/10.1080/13549839.2017.1394830

Hall, P. A., & Taylor, R. C. R. (1996). Political Science and the Three New Institutionalisms. *Political Studies, 44*(5), 936–957. https://doi.org/10.1111/j.1467-9248.1996.tb00343.x

Hansen, M. T., Chesbrough, H. W., Nohria, N., & Sull, D. N. (2000). Networked Incubators. Hothouses of the New Economy. *Harvard Business Review, 78*(5), 74–84, 199. https://doi.org/Article

Hargreaves, T., Hielscher, S., Seyfang, G., & Smith, A. (2013). Grassroots Innovations in Community Energy: The Role of Intermediaries in Niche Development. *Global Environmental Change, 23*(5), 868–880. https://doi.org/10.1016/j.gloenvcha.2013.02.008

Hasanov, M., & Zuidema, C. (2018). The Transformative Power of Self-organization: Towards a Conceptual Framework for Understanding Local Energy Initiatives in The Netherlands. *Energy Research and Social Science, 37*(September 2017), 85–93. https://doi.org/10.1016/j.erss.2017.09.038

Hicks, J., & Ison, N. (2011). Community-Owned Renewable Energy (CRE): Opportunities for Rural Australia. *Rural Society, 20*(3), 244–255.

Hinshelwood, E. (2001). Power to the People: Community-led Wind Energy–obstacles and Opportunities in a South Wales Valley. *Community Development Journal, 36*(2), 95–110.
Hisschemöller, M., & Sioziou, I. (2013). Boundary Organisations for Resource Mobilisation: Enhancing Citizens' Involvement in the Dutch Energy Transition. *Environmental Politics, 22*(5), 792–810. https://doi.org/10.1080/09644016.2013.775724
Hodson, M., Marvin, S., & Bulkeley, H. (2013). The Intermediary Organisation of Low Carbon Cities: A Comparative Analysis of Transitions in Greater London and Greater Manchester. *Urban Studies, 50*(7), 1403–1422. https://doi.org/10.1177/0042098013480967
Hoppe, T., Graf, A., Warbroek, B., Lammers, I., & Lepping, I. (2015). Local Governments Supporting Local Energy Initiatives: Lessons from the Best Practices of Saerbeck (Germany) and Lochem (The Netherlands). *Sustainability, 7*(2), 1900–1931. https://doi.org/10.3390/su7021900
Howells, J. (2006). Intermediation and the Role of Intermediaries in Innovation. *Research Policy, 35*(5), 715–728. https://doi.org/10.1016/j.respol.2006.03.005
Huijben, J. C., & Verbong, G. P. (2013). Breakthrough without subsidies? PV business model experiments in the Netherlands. *Energy Policy, 56*, 362–370.
Kemp, R., Schot, J., & Hoogma, R. (1998). Regime shifts to sustainability through processes of niche formation: The approach of strategic niche management. *Technology Analysis & Strategic Management, 10*(2), 175–198. https://doi.org/10.1080/09537329808524310
Kivimaa, P. (2014). Government-affiliated Intermediary Organisations as Actors in System-Level Transitions. *Research Policy, 43*(8), 1370–1380. https://doi.org/10.1016/j.respol.2014.02.007
Kivimaa, P., Boon, W., Hyysalo, S., Klerkx, L., Rogge, K., Nightingale, P., & Bloom, M. (2017). SWPS 2017–17 (September). *Towards a Typology of Intermediaries in Transitions: A Systematic Review Editorial Assistance, 17.*
Kivimaa, P., & Martiskainen, M. (2017). Innovation, Low Energy Buildings and Intermediaries in Europe: Systematic Case Study Review. *Energy Efficiency, 11*(1), 31–51. https://doi.org/10.1007/s12053-017-9547-y
Kivimaa, P., & Martiskainen, M. (2018). Dynamics of Policy Change and Intermediation: The Arduous Transition Towards Low-Energy Homes in the United Kingdom. *Energy research & social science, 44*, 83–99.
Klerkx, L., & Leeuwis, C. (2009). Establishment and Embedding of Innovation Brokers at Different Innovation System Levels: Insights from the Dutch

Agricultural Sector. *Technological Forecasting and Social Change, 76*(6), 849–860. https://doi.org/10.1016/j.techfore.2008.10.001

Kooij, H.-J., Lagendijk, A., & Oteman, M. (2018). Who Beats the Dutch Tax Department? Tracing 20 Years of Niche–Regime Interactions on Collective Solar PV Production in The Netherlands. *Sustainability, 10*(8), 2807.

Kretzmann, J., & McKnight, J. (1996). Assets-based Community Development. *National Civic Review, 85*(4), 23–29. https://doi.org/10.1002/ncr.4100850405

Lai, W.-H., & Lin, C.-C. (2015). Constructing Business Incubation Service Capabilities for Tenants at Postentrepreneurial Phase. *Journal of Business Research, 68*(11), 2285–2289.

Martiskainen, M. (2017). The Role of Community Leadership in the Development of Grassroots Innovations. *Environmental Innovation and Societal Transitions.* https://doi.org/10.1016/j.eist.2016.05.002

Martiskainen, M., & Kivimaa, P. (2018). Creating Innovative Zero Carbon Homes in the United Kingdom – Intermediaries and Champions in Building Projects. *Environmental Innovation and Societal Transitions, 26*(August 2017), 15–31. https://doi.org/10.1016/j.eist.2017.08.002

Mathie, A., & Cunningham, G. (2003). From Clients to Citizens: Asset-based Community Development as a Strategy for Community-Driven Development. *Development in Practice, 13*(5), 474–486. https://doi.org/10.1080/0961452032000125857

Matschoss, K., & Heiskanen, E. (2017). Making It Experimental in Several Ways: The Work of Intermediaries in Raising the Ambition Level in Local Climate Initiatives. *Journal of Cleaner Production, 169*, 85–93. https://doi.org/10.1016/j.jclepro.2017.03.037

Middlemiss, L., & Parrish, B. D. (2010). Building Capacity for Low-carbon Communities: The Role of Grassroots Initiatives. *Energy Policy, 38*(12), 7559–7566. https://doi.org/10.1016/j.enpol.2009.07.003

Moss, T. (2009). Intermediaries and the Governance of Sociotechnical Networks in Transition. *Environment and Planning A, 41*(6), 1480–1495. https://doi.org/10.1068/a4116

Neumeier, S. (2012). Why do Social Innovations in Rural Development Matter and Should They be Considered More Seriously in Rural Development Research? – Proposal for a Stronger Focus on Social Innovations in Rural Development Research. *Sociologia Ruralis, 52*(1), 48–69. https://doi.org/10.1111/j.1467-9523.2011.00553.x

Ornetzeder, M., & Rohracher, H. (2013). Of Solar Collectors, Wind Power, and Car Sharing: Comparing and Understanding Successful Cases of Grassroots Innovations. *Global Environmental Change, 23*(5), 856–867.

Oteman, M., Wiering, M., & Helderman, J.-K. (2014). The Institutional Space of Community Initiatives for Renewable Energy: A Comparative Case Study of the Netherlands, Germany and Denmark. *Energy, Sustainability and Society, 4*(1), 11. https://doi.org/10.1186/2192-0567-4-11

Parag, Y., Hamilton, J., White, V., & Hogan, B. (2013). Network Approach for Local and Community Governance of Energy: The Case of Oxfordshire. *Energy Policy, 62*, 1064–1077. https://doi.org/10.1016/j.enpol.2013.06.027

Pauwels, C., Clarysse, B., Wright, M., & Van Hove, J. (2016). Understanding a New Generation Incubation Model: The Accelerator. *Technovation, 50–51*, 13–24. https://doi.org/10.1016/j.technovation.2015.09.003

Proka, A., Hisschemöller, M., & Loorbach, D. (2018). Transition Without Conflict? Renewable energy initiatives in the Dutch Energy Transition. *Sustainability, 10*(6), 1721.

Ray, C. (1998). Culture, Intellectual Property and Territorial Rural Development. *Sociologia Ruralis, 38*(1), 3–20. https://doi.org/10.1111/1467-9523.00060

Ray, C. (1999a). Endogenous Development in the Era of Reflexive Modernity. *Journal of Rural Studies, 15*(3), 257–267. https://doi.org/10.1016/S0743-0167(98)00072-2

Ray, C. (1999b). Towards a Meta-Framework of Endogenous Development: Repertoires, Paths, Democracy and Rights. *Sociologia Ruralis, 39*(4), 522–537. https://doi.org/10.1111/1467-9523.00122

Rohracher, H. (2009). Intermediaries and the Governance of Choice: The Case of Green Electricity Labelling. *Environment and Planning A, 41*(8), 2014–2028. https://doi.org/10.1068/a41234

Ruggiero, S., Martiskainen, M., & Onkila, T. (2018). Understanding the Scaling-up of Community Energy Niches Through Strategic Niche Management Theory: Insights from Finland. *Journal of Cleaner Production, 170*, 581–590. https://doi.org/10.1016/j.jclepro.2017.09.144

Ruggiero, S., Onkila, T., & Kuittinen, V. (2014). Realizing the Social Acceptance of Community Renewable Energy: A Process-outcome Analysis of Stakeholder Influence. *Energy Research & Social Science, 4*, 53–63. https://doi.org/10.1016/j.erss.2014.09.001

Schot, J., & Geels, F. W. (2008). Strategic Niche Management and Sustainable Innovation Journeys: Theory, Findings, Research Agenda, and Policy. *Technology Analysis & Strategic Management, 20*(5), 537–554. https://doi.org/10.1080/09537320802292651

Schwencke, A. M. (2017). *Lokale Energiemonitor 2017*. Retrieved from Utrecht: https://www.hieropgewekt.nl/uploads/inline/Lokale_Energie_Monitor_2017_3.pdf

Schwencke, A. M. (2021). *Lokale Energie Monitor 2020*. Retrieved from The Hague: https://www.hieropgewekt.nl/lokale-energie-monitor

Seyfang, G., & Haxeltine, A. (2012). Growing Grassroots Innovations: Exploring the Role of Communitybased Initiatives in Governing Sustainable Energy Transitions. *Environment and Planning C: Government and Policy, 30*(3), 381–400. https://doi.org/10.1068/c10222

Seyfang, G., Hielscher, S., Hargreaves, T., Martiskainen, M., & Smith, A. (2014). A Grassroots Sustainable Energy Niche? Reflections on Community Energy in the UK. *Environmental Innovation and Societal Transitions*. https://doi.org/10.1016/j.eist.2014.04.004

Shucksmith, M. (2000). Endogenous Development, Social Capital and Social Inclusion: perspectives from leader in the UK. *Sociologia Ruralis, 40*(2), 208–218. https://doi.org/10.1111/1467-9523.00143

Smilor, R. W. (1987). Managing the Incubator System: Critical Success Factors to Accelerate New Company Development. *IEEE Transactions on Engineering Management, EM-34*(3), 146–155. https://doi.org/10.1109/TEM.1987.6498875

Smith, A., & Raven, R. (2012). What is protective space? Reconsidering niches in transitions to sustainability. *Research policy, 41*(6), 1025–1036.

Smith, A., Hargreaves, T., Hielscher, S., Martiskainen, M., & Seyfang, G. (2015). Making the Most of Community Energies: Three Perspectives on Grassroots Innovation. *Environment and Planning A*, 0308518X15597908. https://doi.org/10.1177/0308518X15597908

Stewart, J., & Hyysalo, S. (2008). Intermediaries, Users and Social Learning in Technological Innovation. *International Journal of Innovation Management, 12*. https://doi.org/10.1142/S1363919608002035

Süsser, D., Döring, M., & Ratter, B. M. W. (2017). Harvesting Energy: Place and Local Entrepreneurship in Community-based Renewable Energy Transition. *Energy Policy, 101*, 332–341. https://doi.org/10.1016/j.enpol.2016.10.018

Van der Schoor, T., & Scholtens, B. (2015). Power to the People: Local Community Initiatives and the Transition to Sustainable Energy. *Renewable and Sustainable Energy Reviews, 43*, 666–675. https://doi.org/10.1016/j.rser.2014.10.089

Verbong, G., Geels, F. W., & Raven, R. (2008). Multi-niche Analysis of Dynamics and Policies in Dutch Renewable Energy Innovation Journeys (1970–2006): Hype-cycles, Closed Networks and Technology-focused Learning. *Technology Analysis & Strategic Management, 20*(5), 555–573.

Voorberg, W. H., Bekkers, V. J. J. M., & Tummers, L. G. (2014). A Systematic Review of Co-creation and Co-Production: Embarking on the Social Innovation Journey. *Public Management Review*. https://doi.org/10.1080/14719037.2014.930505

Wade, J., Hamilton, J., Eyre, N., & Parag, Y. (2013). Local Energy Governance: Communities and Energy Efficiency Policy. In *ECEEE Summer Study* (pp. 637–648).

Walker, E. T., & McCarthy, J. D. (2010). Legitimacy, Strategy, and Resources in the Survival of Community-Based Organizations. *Social Problems, 57*(3), 315–340. https://doi.org/10.1525/sp.2010.57.3.315

Walker, G., Devine-Wright, P., Hunter, S., High, H., & Evans, B. (2010). Trust and community: Exploring the meanings, contexts and dynamics of community renewable energy. *Energy Policy*, 38(6), 2655–2663.

Warbroek, B., & Hoppe, T. (2017). Modes of Governing and Policy of Local and Regional Governments Supporting Local Low-carbon Energy Initiatives; Exploring the Cases of the Dutch Regions of Overijssel and Fryslân. *Sustainability (Switzerland), 9*(1), 75. https://doi.org/10.3390/su9010075

Warbroek, B., Hoppe, T., Coenen, F., & Bressers, H. (2018). The Role of Intermediaries in Supporting Local Low-Carbon Energy Initiatives. *Sustainability, 10*(7), 2450.

Warbroek, W. D. B. (2019). *The Grassroots Energy Transition: The Success and Governance of Local Low-carbon Energy Initiatives*. PhD University of Twente.

Warbroek, B., Hoppe, T., Bressers, H., & Coenen, F. (2019). Testing the social, organizational, and governance factors for success in local low carbon energy initiatives. *Energy Research & Social Science, 58*, 101269.

Wolsink, M. (2012). The Research Agenda on Social Acceptance of Distributed Generation in Smart Grids: Renewable as Common Pool Resources. *Renewable and Sustainable Energy Reviews, 16*(1), 822–835.

8

Energy Communities Promoting Home Energy Savings: Interventions, Theory and Results

Thomas Hoppe and Frans H. J. M. Coenen

8.1 Introduction

Sustainable energy transitions do not only require the change of energy resources (i.e., from fossil to renewable sources), but also include avoidance of energy spillage and increasing energy efficiency levels (Nejat et al., 2015). Moreover, in light of the energy trilemma, decreasing energy demand has become ever more urgent. However, this is not merely a technological nor an economical challenge. It includes a human factor. In this case behavioral change by final energy consumers, who can potentially consume energy in

T. Hoppe (✉)
Faculty of Technology, Policy and Management,
Delft University of Technology, Delft, The Netherlands
e-mail: t.hoppe@tudelft.nl

F. H. J. M. Coenen
Faculty of Behavioural, Management and Social Sciences,
University of Twente, Enschede, The Netherlands
e-mail: f.h.j.m.coenen@utwente.nl

© The Author(s), under exclusive license to Springer Nature Switzerland AG 2021

F. H. J. M. Coenen, T. Hoppe (eds.), *Renewable Energy Communities and the Low Carbon Energy Transition in Europe*, https://doi.org/10.1007/978-3-030-84440-0_8

more rational and efficient ways. Behavioral determinants of energy users can be targeted with policy measures to induce change or to change the conditions that influence the ways in which energy is consumed. When looking into cities and regions, the realm in which Renewable Energy Communities (RECs) and Citizen Energy Community (CECs) operate, it is in particular households that consume energy and can be targeted with the aim of behavioral change leading to lower volumes in energy demand.[1] This is necessary because globally the residential sector comprises 27 per cent of energy consumption and 17 per cent of total CO_2 equivalent emissions. However, historically households form a target group that is hard to reach or to persuade. There are multiple reasons for this ranging from lack of urgency, knowledge, motivation, and investment capability to reasons that have to do with social norms (Abrahamse et al., 2005; Frederiks et al., 2015). The latter would refer to norms that apply at the neighbourhood level, or more particularly would apply to the social environment to which households and their members belong.

Here energy communities—be they either RECs or CECs—can potentially make a difference. These energy initiatives consist of citizens who jointly seek to produce and consume sustainable energy locally, and many are located in cohesive social communities with high levels of trust (Boon & Dieperink, 2014). In most of these communities, members reject the use of centralized fossil fuel and nuclear energy production, and they embrace decentralized authority and egalitarian values (Hoppe et al., 2019). Energy communities are increasingly considered important players in renewable energy and energy-saving efforts, in terms of raising awareness locally and gaining public support for low-carbon energy and energy savings projects (Coenen et al., 2017). Moreover, energy communities often have local origins and are well embedded in local social systems. This means that they are present in the immediate vicinity of residents. There are certain benefits to this since energy communities already have a social history and reputation with the local community, and can be trusted by the latter. This gives energy communities certain

[1] In this chapter we use the term 'energy communities' to address RECs. However 'energy community' might also cover some CECs since it is sometimes difficult to differentiate between the two.

advantages over others in society—i.e., government and business companies—to stimulate sustainable behaviors like home energy savings among households. This is in part also related to distrust citizens have in government—often culturally determined or based on historical experiences, for example, regarding unfair decision-making processes. Or with business enterprises—often distrusted for only trying to make profits that do not benefit communities and potentially harm the environment (Wolsink, 2007).

In the Clean Energy Package (CEP) RECs and CECs are supposed to focus on providing economic, social and environmental benefits. Economic benefits include: return on investment to members; promotion of local development; investment in public infrastructure; and reduced energy bills—at least for households and other non-professional customers (REScoop.eu and ClientEarth, 2020). Social benefits include: provision of different services (e.g., energy supply, or advice) to members; investment in energy efficiency, energy poverty initiatives; education and training for members or schoolchildren; promotion of energy democracy; and citizen empowerment. Finally, environmental benefits include: increased production of locally produced renewable energy; and greenhouse gas emission reductions (ibid.). However, it is fair to state that in addition, preventing energy spillage by promoting greater energy efficiency and energy savings would also contribute to environmental benefits. From this perspective it is important that energy communities use their agency to promote home energy savings, in particularly among households in their membership base but also among other target groups, and to manage the (renewable) energy systems they own or operate in efficient ways.

The question central to this chapter is: How do energy communities encourage their members (and others; i.e., households) in home energy savings? In addition: To which extent are energy communities effective in doing so? The chapter is structured as follows. In Sect. 8.2 theoretical arguments are provided that are relevant to energy communities undertaking energy savings measures. Section 8.3 addresses research approach and methodology. Section 8.4 presents empirical results and insights. In the concluding section (8.5) the questions are answered, and relevant insights are taken to reflect on the implementation and transposition of the CEP on how energy communities can benefit in terms of strategy with regard to home energy savings.

8.2 Theory: Energy Communities and Interventions to Promote Home Energy Savings

This section consists of three parts. First, it addresses theory on home energy savings by taking a psychological perspective. Second, a range of interventions is addressed. Third, attention is paid to specific ways in which energy communities promote home energy savings.

8.2.1 Part 1: Theory on Home Energy Savings

Households use energy in direct and indirect ways, for different purposes, and in different amounts (Vringer and Blok, 1995). Home energy savings can be referred to in terms of energy conservation behavior, which can be viewed as the effort made to reduce the consumption of energy. This can be achieved in two ways: by using energy more efficiently or by reducing energy demand. Behaviors related to energy conservation (including those with the aim to save energy) can be categorized into curtailment behavior and efficiency behavior. The former concerns ongoing day-to-day actions, such as setting thermostats or switching off lights. Efficiency behavior on the other hand concerns one-time actions to save energy, such as investing in home improvements like thermal insulation or installing energy-efficient appliances (Frederiks et al., 2015).

Frederiks et al. (2015) discern two broad factors that generally explain for energy conservation. These are socio-demographic and psychological factors. Socio-demographic predictors include income, gender and education, but also home-related issues including ownership, size and type of home. Psychological factors on the other hand, include beliefs and attitudes, subjective norms, intentions and perceived behavioral control. Generally speaking, these psychological factors resemble independent variables from the Theory of Planned Behavior (Ajzen, 1991). This theory holds that behavior can be predicted by personal attitudes, subjective norms, perceived behavioral control and behavioral intention. When applied to energy-saving behaviour, a person intends to save a certain amount of energy—which is influenced by attitudes, subjective norms, behavioral control and intention—and then performs actual energy-saving behavior(s) which will (or will not) result in actual energy savings.

8.2.2 Part 2: Interventions Targeting Home Energy-Saving Behaviours

A particular factor that intends to persuade households to engage in home energy-saving behaviors is incentives. In psychology, interventions are considered as actions that are performed to encourage behavioral change. One type of intervention strategy is directed towards activities to modify behavior. Behavioral interventions may be aimed at either voluntary behaviour change, by changing individual knowledge and/or perceptions, or changing the contextual factors (i.e., the pay-off structure) which may determine households' behavioural decisions (Abrahamse et al., 2005). In this chapter, micro-level factors are emphasized (as compared to macro factors that cannot be influenced by energy communities like socio-demographic factors, technological development or economic factors). This holds for activities by energy communities where they try to influence the behavior of their members directly and not through other actors or institutions (e.g., government or public policy). Within these micro-level interventions a distinction can be made between interventions applying either antecedent or consequence strategies. When using an antecedent strategy an agent tries to influence one or more behavioral determinants prior to energy-saving behavior (e.g., providing information with the intention that a household follows up and performs one or more energy-saving behaviors). Examples of actions that use antecedent strategy pertain to commitment, goal setting, information and modelling (i.e., using role models). When using a consequences strategy, an agent tries to influence behavioral determinants of the target group after the occurrence of the energy-saving behavior by providing feedback on performance (e.g., providing feedback on patterns of energy consumption of a given household over the last week). Examples of actions that use consequence strategies are feedback mechanisms and reward systems (ibid.).

Another way to view interventions is in terms of policy (or policy instruments), be it as an instrument implemented by a non-state agent (which differs from government bodies implementing public policy). In comparison to public government energy communities are not steered by elected politicians and do not have a formal mandate based on formal law to formulate and implement policy. Yet most energy communities—in particular renewable energy cooperatives using the seven cooperative principles—apply direct democracy through membership participation, which legitimizes their activities in social ways. Nonetheless, there is no formal legal ground that allows energy communities to exercise power in one way or another over a person, subject matter or territory. Despite this difference, arguably an analogy can be made between actions implemented to by energy communities and the use of public policy by government agents to meet the same goal (i.e., home energy savings).

Policy goals have intended effects (like lowering carbon emissions). For policies to realize the intended impact on society, a large number of people in different situations must make decisions and take actions in line with the intended policy theory to achieve any meaningful impact. This can be achieved only when persons (i.e., target groups) conduct self-initiated actions that contribute to goal achievement, comply with rules or behave in controlled and expected ways in response to incentives that are implemented. In this sense policy emphasizes targeting behavioral determinants addressing individual decisions and actions of certain target groups in society (Schneider & Ingram, 1990). As compared to other target groups end-consumers (including households) are considered difficult to reach (Bressers & Ligteringen, 1997). However, households that also hold membership of energy communities might prove more easy to reach and persuade because they are likely to already have strong (pro-environmental) behavioral attitudes and are exposed to subjective norms in the energy community that favour behavior that will likely encourage energy savings. Nonetheless, actions energy communities use might have the same characteristics as public policy and are therefore appealing to local residents (i.e., households): they provide authority, incentives or

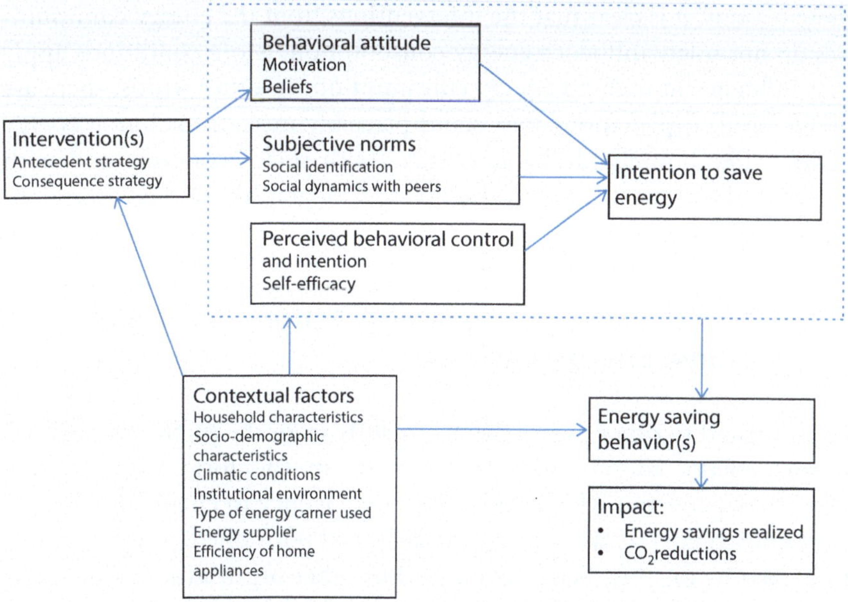

Fig. 8.1 Theoretical framework on interventions regarding home energy savings. (Adapted from: Ajzen (1991); Coenen and Hoppe (2018))

capacity; they use symbolic and hortatory proclamations to influence perceptions or values; or they promote learning to reduce uncertainty (Schneider & Ingram, 1990).

Based on the information provided from both the psychological and policy literature a theoretical framework is developed (see Fig. 8.1). It incorporates insights from different theories and research traditions (mostly Theory of Planned Behavior; Aizen, 2005) and policy evaluation theory (Bressers & Hoogerwerf, 1991; Schneider & Ingram, 1990). In this framework, energy community measures are seen as interventions, and can therefore be found in the box 'intervention(s)'. Here, interventions are expected to directly influence behavioral attitude and subjective norms of households with the aim to encourage them to perform energy-saving behavior(s). The framework, however, also contains the box 'contextual factors'. This is theorized to directly influence energy consumption, next to also influencing most of the other variables in the model. For these reasons, it is clear that the interventions from the energy communities do not solely influence energy consumption. More conditions apply that influence household energy consumption. In other words, there are many (rival) explanations that could plausibly explain for home energy savings, like household characteristics (both socio-economic and with regard to the home as a physical object), demographics and environmental conditions.

8.2.3 Part 3: Energy Community Actions Targeting Home Energy Savings

As compared to other organizations (mainly from the public and private sector) energy communities (most often in the form of renewable energy cooperatives; REScoops) using their agency to encourage citizens (i.e., households) to save energy are fairly well positioned. There are multiple reasons for this. When compared to other organizations (like local government, distribution system operators (DSOs) or energy companies) potentially they can deliver services better for the following number of reasons (Coenen & Hoppe, 2016; Coenen et al., 2017):

- Energy communities are already embedded in social structures, and therefore have close ties with their customer groups, and have direct contacts about energy saving with consumers;
- Energy communities can give personal and tailored assistance to REScoop members to develop personal capacity to save energy;
- Energy communities can organize energy-saving expertise dissemination at the community level, e.g., by organizing workshops, working groups or setting up an 'energy library';
- Energy communities can build home energy saving expertise through gaining critical mass;
- Energy communities can cooperate better with local stakeholders;
- Energy communities can raise awareness among both the larger community and individual members to stress the importance of energy saving; because of their social embeddedness in local communities they likely reach out better to target groups than other agents would;

- Energy communities can set energy saving as a social norm within the community;
- Energy communities are locally viewed as a reliable partner to give advice, supply energy systems and appliances and make people more willing to take home energy saving investment risks;
- Energy communities can tailor home energy saving measures to where it is effective, while also addressing related social issues like energy poverty and justice;
- Energy communities can define and distribute the available capacity of renewable energy as a common resource;
- Energy communities can deal with NIMBY problems related to aspects that have to do with siting (renewable) energy plants while balancing spatial, social, economic and environmental interests.

Energy communities pay attention to home energy saving in various ways. This can be done, for example, by raising awareness, providing education and training to households and advisers, but also by providing support in audits and implementation processes. A simplified way to see how this works in practice with a focus on how an energy cooperative deals with this and takes measures to convince its members to start saving energy. The way in which it implements energy saving activities varies from providing information and advice to actively guiding households and supporting them in implementing saving measures (Coenen & Hoppe, 2016).

Energy communities can influence the behaviors of their members in different ways. To create more clarity conceptually, we will address what energy community membership entails vis-à-vis general membership, actions and events that energy communities organize. Therefore, three (overlapping) (sub)sets of energy community membership are discerned. In this classification general membership is the first way an energy community influences its members. This concerns an overlapping set of the way energy communities influence their members. Within this set there is the engagement of members in the energy community, for instance by organizing meetings, via financial participation or via participation in dedicated actions or measures organized by the energy community

Fig. 8.2 Venn diagram of energy community membership (here applied to REScoops), discerning between general membership, engagement in REScoop activities and participation in specific dedicated REScoop actions. (From: Coenen & Hoppe, 2018)

(Coenen & Hoppe, 2018). Figure 8.2 presents a graphical overview of the influence spheres an energy community has on its members.

To link members' home energy savings behavior to the assumed influence of energy community actions we discern between specific and unspecified measures in relation to the member's (actual) behaviour. Unspecified measures entail the general presumed influence of being a member and being (indirectly) exposed to energy community actions and information. On the other hand, specific dedicated measures concern interventions for which information is available on which members participate in these specific actions or are exposed to them in another way (by becoming a target group for a specific intervention) (ibid.).

Here, it is assumed that the mechanism of energy communities influencing their members to attain certain goals works as follows. Membership potentially influences energy saving behaviors for a number of reasons. Becoming a member (and/or customer) can be seen as making an informed choice; in other words, one chooses deliberately to engage in using green energy. The reason to become a member can be motivated by environmental or sustainability concerns or by pragmatic financial or technical reasons, like the expectation to receive better service provision or more comfort. If one obtains membership, one receives information on the importance and ways to save energy. This could mean that the information level of the energy community members on the importance of renewable energy and possibilities to save energy increases after obtaining membership which could lead to a higher knowledge level (on renewable energy and energy saving options). However, more information or awareness does not automatically mean that one also engages in actions to attain a certain goal (like saving a certain amount of energy). We assume that it is more easy for energy communities to influence members who are more financially and actively engaged in their energy community, for instance because they hold shares in their energy community or visit meetings it organizes. This is a particular subset of energy community members; i.e., the subset of engaged members (ibid.).

8.3 Research Approach and Methodology

The EU's Horizon 2020 project 'REScoop Plus' (2016–2018) (Chalkiadakis et al., 2018; Coenen et al. 2017) addressed energy communities using their agency to encourage home energy savings. This project researched and sought to improve energy savings strategies as an activity for renewable energy cooperatives (REScoops) across Europe. The research activities of the project involved two work packages: one addressing statistical analysis of energy consumption longitudinal data sets of REScoops, and another addressing behavioral aspects of REScoop incentives targeting home energy savings. The latter addressed the question, *"What behavioral and social aspects influence energy savings by households, including REScoop members?"*, with the aim of empirically testing the claim whether REScoops can be considered effective agents in persuading households in achieving home energy savings.

The project included a wide set of research methodologies: exploratory analysis of the ways in which REScoops use their agency, to map the incentives, measures, tools and approaches they use, and conducting surveys to collect data that would allow for statistical analysis. The 'modus operandi' methodological approach was used to analyse whether REScoops' agency would explain for behavioral change of households or not. Two rounds of surveys were performed among REScoop members and others. First, in the spring and summer of 2017 a survey was conducted among six REScoops in five EU states, with a total response of 10,585 households. Second, in the spring and summer of 2018 a survey was conducted among seven REScoops in six EU nations (N = 7556). Whereas the 2017 survey focused on general REScoop issues and home energy savings, the 2018 survey paid more intention to the implementation of a number of dedicated REScoop measures (i.e., specific interventions). The results presented in this chapter are mostly based on surveys among REScoop members, non-member clients (consuming energy supplied by REScoops), or persons otherwise connected to the REScoop community. The REScoops that were part of the research project included: Coopernico (Portugal), ènostra (Italy), Ecopower (Belgium), Enercoop (France), EBO (Denmark), SEV (Italy) and SOM energia (Spain) (ibid.).

It should be noted here that the study on analyzing the actual consumption data focused on electricity consumption (excluding gas and other sources needed for heating of homes and tap water). This was for reason of availability of data. Section 8.5.2's behavioral analysis focused on both electricity and energy sources used for in-home heating. In the next section of this chapter results from REScoop Plus are presented. However, to generate a more up-to-date depiction of the state of affairs on how energy communities use their agency to achieve household energy savings an additional in-depth case is presented (the Netherlands) that goes beyond merely focusing on energy communities as implementing agents and households as target group. The case study also pays attention to the structural and institutional environment as well as to the role of intermediary agents (also with regard to empowering of energy communities).

8.4 Results

8.5 Overview of Interventions

In this section the question, "How do energy communities encourage their members (and other households) in home energy savings?" is addressed. To answer this question an overview was established of the variety of action and interventions energy communities use. When looking internationally at energy-saving practices of energy cooperatives in Europe, an overview can be presented, in which measures can be classified into the dimensions of energy behavior (with the extremes of the dimension as either curtailment of efficiency behavior) and type of strategy (either antecedent or consequence strategy). This results in a 2×2 matrix with four quadrants in which the measures are positioned (See Fig. 8.3). The measures observed stem from different overview studies including but not limited to the EU-H2020 'REScoop Plus' project (Coenen & Hoppe, 2016; Schwencke, 2021). An overview of specific dedicated measures used by energy communities is presented in Table 8.1.

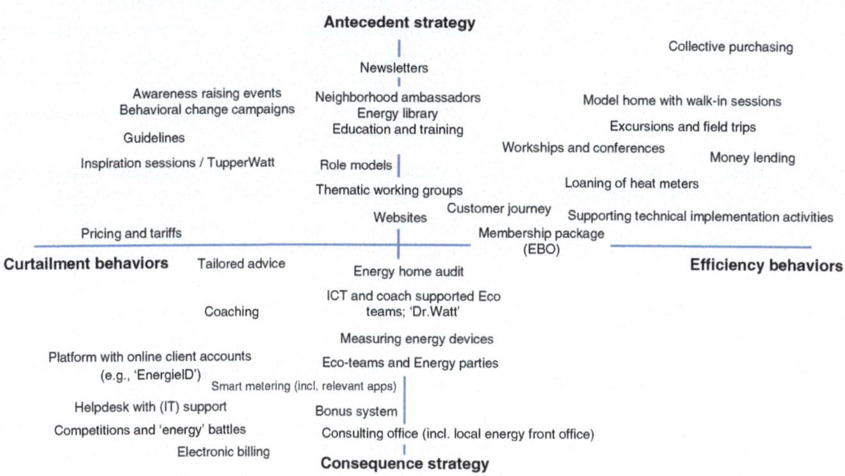

Fig. 8.3 Classification of measures implemented by energy communities to persuade households to save energy

Table 8.1 Overview of specific dedicated measures used by energy communities

Measure	Implemented by (energy cooperative, country)	Description
Dr. Watt	Enercoop (France)	An online tool that goes along with an offline training course to help consumers make a self-diagnosis of their electricity consumption. The approach seeks to make households more aware and to increase understanding into home electricity consumption, but also provides tailored advice.
TupperWatt	Enercoop (France)	Meetings organized for households that want to be more involved with energy community activities and that put citizens at the centre of energy issues. This type of event—Inspired by "Tupperware parties"—fits the general communication strategy of Enercoop: Not too much advertising, creating social links within the community while sharing experiences.
EnergieID	Ecopower (Flanders – Belgium)	A SaaS-platform (i.e., "software as a service") to support households to understand and manage their energy consumption as well renewable energy production (via solar panels). Customers sign up with an account on EnergieID and on a monthly basis they fill in their energy consumption data. Then together with the help desk service of Ecopower the energy bills and energy consumption are analyzed and discussed with customers (including Ecopower members); either by phone or by email.
DH package	EBO (Denmark)	'Pakkeløsning' in Danish—a conversion package for home owners to switch from a gas grid connection to a (sustainable) DH system grid connection. It includes four steps: (1) A home visit and an agreement of where the district heating unit is going to be installed; (2) Building and connecting a heat supply line to the consumer's home and restoration of the garden; (3) A removal of the consumer's existing heating system; (4) A delivery and an installation of a new district heating unit. Before the measures take place unburdening of the home owner takes place. Afterwards performance of the installed DH system package is monitored periodically.

(*continued*)

Table 8.1 (continued)

Measure	Implemented by (energy cooperative, country)	Description
Customer journey	HIER opgewekt / Hoom (the Netherlands)	The customer journey concerns a stepwise approach to support home owners to make one's home more energy-efficient. It starts with awareness and inspiration, and includes checking whether the technical energy measure(s) installed meets the expectations set. This consists of six steps: Orientation, advice, quotation, execution, control and finally inspiring others. Trust is essential throughout the customer journey process. Home owners are looking for a reliable partner who will act in their best interest. Energy cooperatives take up this role and guide residents in making their homes more energy-efficient.
Info Energeia	SOM Energia (Spain)	A personalized energy awareness service, which provides information to households consumers via two channels. First, via a monthly report (in PDF) with benchmarking of a household's energy consumption against peers, and with useful tips. Second, via an online customer portal where the monthly reports and extended information are made available (including smart metering measurements).
Energy ambassadors	Hoom (the Netherlands)	A volunteer who conducts conversations with neighbourhood residents and guides them in the process of making their homes more energy efficient. Energy coaches know what making homes more energy efficient entails, both in terms of construction and installation technology. Energy ambassadors inspire and inform local residents based on their own experiences.

As can be observed in Fig. 8.3 there is a wide range of measures energy communities use. The figure reveals that the majority of measures uses antecedent strategy rather than consequence strategy and that curtailment behavior appears to be targeted more than efficiency behavior. Moreover, in terms of actions in terms of 'policy instruments' it can be observed that the majority of measures can be seen as capacity tools with the aim of informing target groups about the benefits of energy savings and to prepare how to engage in energy saving behaviors. Examples include the use of energy ambassadors, awareness raising events, inspiration sessions and using mock homes including state of the art energy efficient technology as a role model. Incentive tools (like rewards or competitions) are also observed but appear less frequently. The measures observed also include technological tools like energy communities lending infrared heating meters to households enabling them to observe thermal bridges in their homes, smart meters to measure and provide feedback on energy consumption, and ICT interfaces to support energy service delivery—including tariffs and billing—and information provision to households (as 'clients'; and energy community members at the same time). The mapping exercise also revealed integrated measures that include a multitude of actions, and contain both antecedent and consequence strategy. Examples include the 'Dr. Watt' training programme of French REScoop Enercoop, which includes elements of eco-teams, role models, training and education, tailored advice, smart metering and ICT support. Hence, it includes both capacity and incentive tools, yet also uses technological tools in support (i.e., ICT, home metering equipment and a smart meter).

8.5.1 Use of Energy-Saving Measures by Energy Community Members

Second, an overview is created on which actions are used and in what frequency are used by energy community members at the energy cooperative LochemEnergie in the Netherlands (See Fig. 8.4) (Hoppe & Coenen, 2016). It reveals a wide variety of actions, mostly focusing on antecedent strategy. The chart reveals that there are no less than seven actions in which about half of the members or more participate.

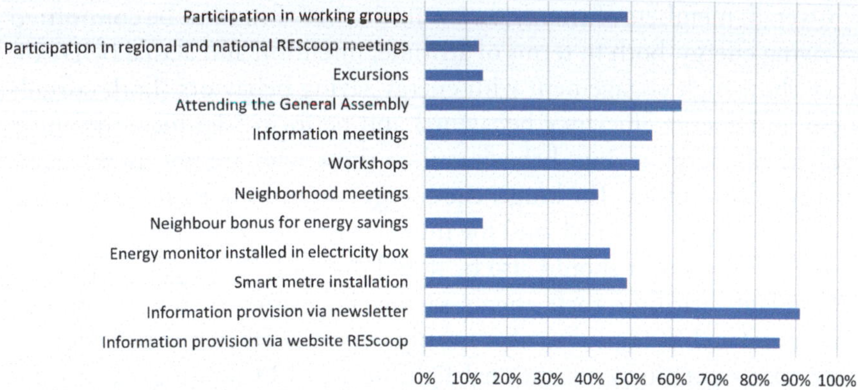

Fig. 8.4 Overview of members participating in dedicated energy savings actions implemented by energy cooperative LochemEnergie (in relative frequency of total membership)

8.5.2 Behavioral Analysis

The 'REScoop Plus' project surveys indicate that energy community members undertake many (individual) energy savings actions like lowering the thermostat or taking shorter showers. Only a fairly small part of those respondents, however, indicates that undertaking (individual) energy savings actions can be attributed to a REScoop (e.g., 18 per cent of the respondents from Enercoop; and 36 per cent of respondents from Ecopower) (Hoppe et al., 2019). For energy curtailment behaviors this is considerably less (15–17 per cent) than for energy efficiency behaviors (20–30 per cent). Between 36 per cent (Enercoop) and 64 per cent (Ecopower) of the surveyed energy community members indicate to using less energy savings after obtaining energy community membership. And 44 per cent of the respondents argued that energy community membership has contributed to household energy savings. Respondents indicate average energy savings in the range of 4–6 per cent. Of those who measured their energy consumption, about 21–22 per cent indicate to have saved at least 10 per cent energy, and between 9 per cent and 10 per cent indicates to have saved at least 20 per cent (Coenen & Hoppe, 2018).

Overall, members of energy communities were found to be committed to saving energy, both in terms of attitude, intention and actual behavior. They show high engagement with energy saving behaviors (both curtailment and energy efficiency behaviors) and perform (significantly) more individual energy savings behaviors than those who are not members of energy communities. The longer the energy community membership, the more knowledge is gained, and the more energy saving behaviors are performed. However, it should also be noted that only 5–20 per cent of the respondents attribute their energy savings behaviors to energy community membership (and if so this mostly concerns efficiency behavior, in particular switching conventional lighting to LED lighting). It can be argued though that regarding curtailment behaviors energy community members arguably were already engaged in energy saving behaviors prior to becoming energy community members. Other energy community factors that show a significant and positive statistical relation to energy savings (and intention to save energy) concern visiting of energy community meetings (or workshops), and integrated (i.e., combination of) measures like EnergieID (Ecopower), Dr. Watt (Enercoop) and InfoEnergia (SOM Energia). All three combine ICT tools with antecendent and consequence strategies, including feedback on actual consumption (ibid.).

Research was also conducted on the influence of rival factors including psychological predictors (e.g., goal-setting), social predictors (in particular social environment), knowledge level, demographics and household characteristics (i.e., education level, home size, and ownership). In multivariate regression analyses using three indicators of energy savings the most significant predictor turned out to be psychological factors. REScoop measures only modestly contributed to the explained variation of the models. However, the analysis also found strong statistical covariation between REScoop social/organizational factors and psychological predictors, stressing the importance of psychological predictors, which can potentially be targeted by cooperatives in their strategies to persuade their members to conserve energy (Hoppe et al., 2019). The study's results also support the claims that interventions are more effective when tailored to the target population (Abrahamse et al., 2005) and when combining informational and structural strategies with antecedent and consequence strategies (Gardner & Stern, 1996).

8.5.3 Effects of Dedicated Energy Saving Measures Implemented by Energy Communities

When looking into the use of specific integrated measures (including both antecedent and consequence strategy like the Dr. Watt intervention by Enercoop) implementation can be considered as fairly effective, with considerable energy savings as compared to the start situation. In the 2017 survey a number of specific energy measures and tools implemented by REScoops (i.e., Dr. Watt training sessions, personal advice, or EnergieID) were found to significantly and positively relate to energy savings (since becoming a REScoop member). Moreover, users were generally satisfied with them. For EnergieID users also indicated increased importance and contribution to energy savings. Increasing portions of the respondents indicated use (e.g., EnergieID: from 20 per cent in 2017 to 30 per cent in 2018 at Ecopower; Dr. Watt: from 3 per cent of Enercoop members to 37 per cent in 2018). Results from the 2018 survey revealed that specific measures using platforms (along with related informational actions) were found to statistically correlate positively to reported energy savings, whereas sole informational actions (e.g., TupperWatt, or saving tips on the energy saving Wiki) only influenced intention to save energy, but not energy savings itself (Coenen & Hoppe, 2018).

Longitudinal time series trend analysis revealed a number of important findings (Sifakis et al., 2018). The key finding is that implementing energy efficiency interventions of various types, such as technical support, special tariffs, energy generation schemes and smart meters, leads to substantial energy reductions of more than 10 per cent, cumulatively (Sifakis et al., 2020). More specifically, joining a REScoop was found to lead to a more than 20 per cent reduction in electricity consumption. And installing solar panels on one's home reduces REScoop members' electricity demand by more than 45 per cent, with those having solar panels installed at home consuming nearly three times less grid-supplied electricity than those who do not have solar panels installed at home. At Ecopower no less than 43 per cent of the respondents were found to be prosumers, generating their own green power, locally. The share of

Ecopower members having installed solar panels at home did also increase sharply over recent years, encouraged by Ecopower's agency (Sifakis et al., 2020). Further, energy efficiency interventions of various kinds, such as technical support, special tariffs, energy generation schemes and installing smart meters, statistically correlate (positively) to substantial reductions in energy consumption: i.e., those who register with EnergieID save 10 per cent in energy consumption; those who partake in Dr. Watt training sessions at Enercoop were found to consume 13 per cent less electricity than others who did not partake; and those who had smart meters installed were found to consume over 26 per cent less electricity. The longitudinal data analysis results should however be interpreted with caution as limitations in the data collection (mostly due to challenges to the availability of reliable data) caused the research to only modestly address (internal and external) validity issues. Installing a solar panel system might for instance create a distorted image with regard to the influence of smart meter installation to total household electricity consumption.

8.5.4 Support Structures and Energy Communities Cooperating with Government

To get an impression of which interventions are used, a case study of the Netherlands was used to provide more in-depth insights into how this works at the national level. A 2020 survey reveals that almost half of the energy cooperatives indicate that they are engaged in promoting energy saving measures. Energy cooperatives are active in providing information via their website, newsletters and district or via regional meetings. A number of cooperatives also use energy coaches. Energy coaches are local residents that have been trained in home energy affairs, and who make an inventory of wishes, and answer questions about measures that lead to energy savings. The advantage of this approach over business initiatives is that energy coaches are more familiar with the local situation and therefore have more confidence. Some energy cooperatives offer an integrated approach: with a local approach, a personal approach to households, with a network of local entrepreneurs and active cooperative members who act as energy coach and neighbourhood ambassador, and with guidance during implementation (Schwencke, 2021).

An important observation made is that energy cooperatives rarely operate alone. For example, they often collaborate with local government (in the implementation of local 'energy front offices', giving advice, coaching and cooperating in local projects) and other social initiatives that are also involved in energy saving, such as 'Buurkracht' (which, incidentally, was once founded as an initiative from a distributed system operator). Research by 'HIER opgewekt' shows that when municipalities work together with energy cooperatives, they reach more households than without (ibid.). Another form of cooperation between energy cooperatives and governments is by making use of a public (subsidy) scheme ('Energy Consumption Reduction Scheme') aimed at encouraging households to take saving measures. This finances the training of energy coaches or the organization of residents' actions with small savings measures—such as with an 'energy box'. Ever since local governments and energy cooperatives started to collaborate, both the number of trained energy coaches and the number of coaching conversations (with local households) have increased. For this reason it is argued that local governments view energy cooperatives as a good cooperation partner for reaching households (ibid.).

Another observation is that a national energy cooperative ('Hoom') has been established by the cooperative movement that specifically functions as an umbrella and as a platform to support other energy cooperatives to build capacity to properly prepare and perform their energy saving activities. 'Hoom' shares knowledge, makes expertise available, trains energy coaches, develops smart tools and offers practical support. However, in line with its vision 'Hoom' assumes that local cooperatives follow their own approach, in collaboration with local partners and implementers. To support local initiatives, 'Hoom' has developed the "Cooperative Energy Desk", which supports local approaches from a national organization with IT support and additional services. In 2020, seventy-seven local energy cooperatives were affiliated to 'Hoom' (out of a total of 623 energy cooperatives nationwide). An inventory among 'Hoom' members showed that the total number of realized energy saving measures (such as thermal insulation, but also replacement of heating elements) has recently increased sharply (ibid.).

8.5.5 Conclusion

This chapter started with raising two research questions. First: How do energy communities encourage their members (and other households) in home energy savings? And second: To which extent are energy communities effective in doing so?

Three ways were presented in this chapter in which energy communities can influence households in general, and more specifically their members. First, via the social structure and norms that pertain to energy community membership, assuming that households obtain (or maintain) energy community membership. Second, via active engagement of households but in a general sense (e.g., reaching out to them by organizing energy community meetings). Or third, by implementing certain dedicated actions and measures to persuade households to save energy (e.g., via energy ambassadors, providing tailored advice, or using a team approach that also features feedback on home energy consumption performance). Results from empirical studies show that all three are used by energy communities.

Regarding membership it can be argued that by becoming an energy community one is influenced by its social environment to develop a certain attitude that contributes to setting intentions to save energy. Between 36 per cent (Enercoop) and 64 per cent (Ecopower) of the surveyed energy community members indicate to use less energy savings after obtaining energy community membership. And 44 per cent of the respondents argue that energy community membership has contributed to household energy savings. According to the 'REScoop plus' project survey energy savings are on average in the range of 4–6 per cent, and according to longitudinal trend analysis even range up to more than 10 per cent. Results also show that the longer one holds energy community membership the more one engages in energy savings behaviors and the more one actually succeeds in saving energy. Regarding engagement, a number of indicators (like attending meetings organized by energy communities) were found to have a significant positive statistical relationship to energy savings reported by energy community members. With regard to the implementation of dedicated energy community actions and measures several strategies were found to have a significant and positive statistical relationship to energy saving behaviors and achieving energy saving, and were also found to be highly appreciated. These in particular concern the Dr. Watt program, Smart Meter installation, and EnergieID and InfoEnergia. Energy communities encouraging their members to install solar panels and become 'prosumers' was also found to have an impact (Sifakis et al., 2020). Moreover, the in-depth case study revealed that using energy ambassadors and local energy front offices ('energieloket' in Dutch) also seems to work effectively as they are well-practiced in this work and are valued by different stakeholders (Schwencke, 2021).

In sum, it looks like the three forms of engagement (i.e., membership itself, engagement activities, and the use of specific measures) contribute in a positive way to households' energy saving intentions, behavior and eventually energy savings. One may argue, though, that energy community membership and engagement strategy are necessary conditions that influence household energy saving behavior. Although some results indicate that these factors alone can already encourage household energy saving behaviors, results of the analysis of dedicated measure implementation reveal that they can trigger and reinforce these conditions. In summary, membership, engagement activities and specific measures appear to reinforce each other, and are, arguably, jointly best possible to trigger energy-saving behaviours among REScoop members. To answer the second question: energy communities are well positioned and fairly capable of encouraging household energy savings (taking into account that many households, and in particular those having energy community membership, already hold predispositions that favor energy saving behaviors).

Insights from the case study of the Netherlands revealed that energy communities do not stand alone in encouraging household energy savings. First, the energy community movement managed to establish a dedicated cooperative ('Hoom') to serve as an intermediary agent to support individual energy communities (most of them local energy cooperatives) in becoming knowledgeable about energy saving interventions, offering a dedicated toolbox and building capacities. Recent developments indicate that energy cooperatives are increasingly using the services 'Hoom' offers, while at the same taking more energy-savings actions (mostly by using energy ambassadors, by offering 'energy box' kits to households, and by using the customer journey approach to persuade and support households), while managing to get more thermal insulation and other energy saving equipment installed. On the other hand, energy communities encouraging home energy savings also appear to increasingly cooperate with government. First, by using a national government funded scheme to finance the training of energy ambassadors. And second, by cooperating with local government in shaping and implementing local energy front offices and reaching out to citizens that are difficult to reach for local governments themselves. These can be viewed as energy communities 'co-producing' local climate policy and implicate energy communities operating on arm's length of government.

Transposition and implementation of CECs and related issues from the Internal Electricity Market Directive into national and decentralized legislation could potentially cover these kind of policy schemes at the national level that address capacity building and co-production of energy saving services by energy communities. This would arguably encourage and secure ongoing practices of energy community agency in energy savings in front-runner countries, and disseminate a potential institutional framework to countries in which energy communities still have to start up activities with regard to encouraging household energy savings.

References

Abrahamse, W., Steg, L., Vlek, G., & Rothengatter, T. (2005). A Review of Intervention Studies Aimed at Household Energy Conservation. *Journal of Environmental Psychology, 25*, 273–291.

Ajzen, I. (1991). The Theory of Planned Behavior. *Organizational Behavior and Human Decision Processes, 50*(2), 179–211.

Boon, F. P., & Dieperink, C. (2014). Local Civil Society Based Renewable Energy Organisations in the Netherlands: Exploring the Factors That Stimulate Their Emergence and Development. *Energy Policy, 69*(0), 297–307. https://doi.org/10.1016/j.enpol.2014.01.046

Bressers, J., & Ligteringen, J. (1997). What to Do with Non 'accessible' Target Groups: Policy Strategies for Sustainable Consumption. CSTM nr. 66. Centrum voor Schone Technologie en Milieubeleid. Enschede. pp. 1–23.

Bressers, J. T. A., & Hoogerwerf, A. (Eds.). (1991). *Beleidsevaluatie.* Samsom H.D. Tjeenk Willink.

Chalkiadakis, G., Akasiadis, C., Savvakis, N., Tsoutsos, T., Hoppe, T., & Coenen, F. (2018). Providing a Scientific Arm to Renewable Energy Cooperatives. In *The Role of Exergy in Energy and the Environment* (pp. 717–731). Springer.

ClientEarth, R. e. a. (2020). *Energy Communities under the Clean Energy Package; Transposition Guidance.* Retrieved from Brussels: https://uploads.strikinglycdn.com/files/48701cfd-f397-4903-9d36-1fba162223f4/Energy%20Communities%20Transposition%20Guidance.pdf

Coenen, F., & Hoppe, T. (2016). *D3.1 Report on Specific Tools of Supplying REScoops in Europe* (Deliverable D3.1). Retrieved from Enschede/Delft:

Coenen, F., & Hoppe, T. (2018). *D3. 4–Effectiveness Report 2.* Retrieved from Enschede: University of Twente.

Coenen, F. H., Hoppe, T., Chalkiadakis, G., Tsoutsos, T., & Akasiadis, C. (2017). Exploring Energy Saving Policy Measures by Renewable Energy Supplying Cooperatives (REScoops). In *ECEEE 2017 Summer Study on Energy Efficiency: Consumption, Efficiency and Limits.*

Frederiks, E. R., Stenner, K., & Hobman, E. V. (2015). The Socio-demographic and Psychological Predictors of Residential Energy Consumption: A Comprehensive Review. *Energies, 8*(1), 573–609.

Gardner, G. T., & Stern, P. C. (1996). *Environmental problems and human behavior.* Allyn & Bacon.

Hoppe, T., & Coenen, F. (2016). *Exploring Interventions and Tools Used by REScoops to Lower Householders' Energy Consumption and Stimulate Investment in RES Projects.* Paper Presented at the NIG Annual Work Conference, Antwerp. November 24, 2016.

Hoppe, T., Coenen, F. H., & Bekendam, M. T. (2019). Renewable Energy Cooperatives as a Stimulating Factor in Household Energy Savings. *Energies, 12*(7), 1188.

Nejat, P., Jomehzadeh, F., Taheri, M. M., Gohari, M., & Majid, M. Z. A. (2015). A Global Review of Energy Consumption, CO_2 Emissions and Policy in the Residential Sector (with an Overview of the Top Ten CO_2 Emitting Countries). *Renewable and Sustainable Energy Reviews, 43*, 843–862.

Schneider, A., & Ingram, H. (1990). Behavioral Assumptions of Policy Tools. *The Journal of Politics, 52*(02), 510–529.

Schwencke, A. M. (2021). *Lokale Energie Monitor 2020.* Retrieved from The Hague: https://www.hieropgewekt.nl/lokale-energie-monitor

Sifakis, N., Daras, T., & Tsoutsos, T. (2020). How Much Energy Efficient are Renewable Energy Sources Cooperatives' Initiatives? *Energies, 13*(5), 1136.

Sifakis, N., Savvakis, N., Daras, T., & Tsoutsos, T. (2018). *D2.4 – Final Data Statistical Analysis*. Retrieved from Chania, Greece.

Vringer, K., & Blok, K. (1995). The Direct and Indirect Energy Requirements of Households in the Netherlands. *Energy Policy, 23*(10), 893–910.

Wolsink, M. (2007). Planning of Renewables Schemes: Deliberative and Fair Decision-making on Landscape Issues Instead of Reproachful Accusations of Non-cooperation. *Energy Policy, 35*(5), 2692–2704. https://doi.org/10.1016/j.enpol.2006.12.002

9

Creating an Enabling Policy Framework for Inclusive Energy Communities: A Gender Perspective

Mariëlle Feenstra and Florian Hanke

9.1 Introducing the Gender–Energy Nexus in Just Energy Transitions

The current context and future consequences of climate change require a change in how energy is produced and consumed. However, to secure a just and inclusive energy transition, there is a need to understand both the effects of a low-carbon energy transition on different policy levels and groups in society, and how people can engage in such a process on an equal basis. Social inequalities and energy justice are two key emerging streams in the energy policy literature, with little but recently growing attention to gender equality (Feenstra & Özerol, 2021). Social

M. Feenstra (✉)
BMS-CSTM, University of Twente, Enschede, The Netherlands
e-mail: m.h.feenstra@utwente.nl

F. Hanke
European University Viadrina, Frankfurt/Oder, Germany
e-mail: hanke@europa-uni.de

© The Author(s), under exclusive license to Springer Nature Switzerland AG 2021

F. H. J. M. Coenen, T. Hoppe (eds.), *Renewable Energy Communities and the Low Carbon Energy Transition in Europe*, https://doi.org/10.1007/978-3-030-84440-0_9

inequalities are the heart of social justice theories and shaping the gender lens of feminist theories. Within the gender–energy nexus, three roles of actors in the energy system are identified: consumers, producers and decision-makers (Clancy & Feenstra, 2019). Within energy communities, prosumers, energy consumers that produce a part of the energy consumed, are combining all three roles as change agents in the energy transition. Female prosumers can play a crucial role as agents of change towards a just energy transition in their communities. Additionally, energy communities have the potential to empower their members to participate in energy transitions.

This chapter dives deeper into the mutual engagement of female prosumers and energy communities and the potential of energy communities to contribute to a more just and inclusive energy transition. Capabilities embedded in gender inequality of societies might limit female prosumers' potential to use their agency to participate in energy transition processes. Policy interventions can contribute to use the potential of female prosumers and to foster just transitions. Women's role as change agents in claiming energy justice and overcoming social inequalities in energy access is embedded in the Sustainable Development Goals (SDGs) that address gender equality (SDG5), eradicate poverty (SDG1), and access to affordable and clean energy (SDG7). Despite the global commitment, we observe that the enforcement of national governments is limited, creating a lack of implementation of inclusive energy transitions at the local level.

The current context and future consequences of climate change necessitate that we accelerate towards a decarbonised energy future by replacing fossil energy with renewable energy sources (RES) and reducing energy consumption. This challenge comes in addition to making access to sustainable energy universal in both the Global South and Global North to ensure development and social justice. The Global South and Global North are concepts that are not used in a strict geographic sense but rather in a political economic one of large disparities in wealth and political instability. Global commitments such as the Paris Agreement, Sustainable Energy for All (SE4ALL) and the Sustainable Development Goals (SDG); SDG#7—'clean and affordable energy for all' and SDG#13 on 'climate action', are important frameworks for government policy implementation in the energy transition. The energy transition is often presented as depending on technological innovation and market mechanisms. Replacing fossil fuels like coal with RES in electricity productions is highly beneficial, reduces emissions and ensures continued energy security. However, a successful low-carbon energy transition requires equal attention to the political and social dimensions. There is an urgent need to recognise both the effects of the global energy transition on different policy levels and societal groups and how people can engage in the process on an equal basis. Furthermore, we need to understand how this knowledge can be translated into policies, incentives and communication (Grübler et al., 2016). A socially inclusive energy transition is necessary to ensure social acceptance of new energy technologies (Sovacool et al., 2017) and to avoid the detrimental consequences of the energy transition from taking place (Fuller, 2019; Standal et al., 2018; Clancy et al., 2017).

Despite the urgency of the situation, there is a policy gap in relation to gender in the low-carbon energy transition. The research question of this chapter is: what is the potential of energy communities to contribute to a more just and inclusive energy transition and how can European policy support them? On the policy level in the Global North, especially in the EU, there has been an emerging focus on engaging consumers to foster sustainable energy consumption and to empower people to become managers of their energy needs focusing on energy behavioural change (Standal et al., 2019). The Clean Energy For All Europeans (European Commission, 2019) package of legislation creates a first step towards an enabling policy framework for a more inclusive energy transition of the European Union. Considering the free-market economy and the non-discrimination law, the assumption is that industrial countries have gender-neutral energy policies that benefit both men and women equally (Standal et al., 2018; Clancy et al., 2017; Clancy & Röhr, 2003). Thus, national energy policies do not consider that women and men have different opportunities and ways of engaging with energy on policy, research and corporate levels (Standal et al., 2018; Pearl-Martinez & Stephens, 2016; Clancy & Röhr, 2003). An explanation for the gender-blindness of energy policy-makers is the lack of awareness due to limited available gender disaggregated data (Feenstra & Clancy, 2020). RECs (renewable energy communities) that are operating at the local level in close connection with the communities and the energy users, might be more aware of the needs of energy users and the inequalities in access.

This assumption is researched in the Global South in which local energy initiatives are proven to be more aware of inequal energy access (Standal et al., 2018). In the policy context of the Global South, the gender–energy nexus has gained more traction, stimulated by donor organisations and financial support by international institutions like the UN and World Bank to stimulate compliance to international agreements (Standal et al., 2018). However, women have mainly been presented under policy agendas that focus on women as victims (without agency) or neoliberalist ideas in which energy access is instrumental in transforming women to become productive citizens securing economic growth and well-being to their families, communities and the national level (Listo, 2018; Standal et al., 2018). These policy logics, including the gender blind/neutral logic of the Global North, do not treat the underlying mechanisms of discrimination that reinforce gender inequality in the political economy of energy and therefore provide an inadequate understanding of women's roles and resources in relation to energy. A just energy transition can be engendered by highlighting women's agency in the energy system. This chapter aims to contribute to the energy justice and just transitions' debate, and the growing body of literature pertaining to the gender–energy nexus in the light of energy communities as an emerging actor in the energy system.

Given that energy access, consumption and services globally are gendered (Clancy & Feenstra, 2019; Standal et al., 2018; Fraune, 2015; Bell et al., 2015; Clancy & Röhr, 2003), establishing an energy system that reflects gender differences, and that is aware of gendered relations in society, is essential for a just energy transition. Acknowledging the socially constructed and dynamic nature of gender relations, energy policy should acknowledge cultural differences in their social context and should contribute to the overall aim of gender equality (Feenstra & Özerol, 2021). Engendering energy policy is defined as the process that aims to create a gender-aware energy policy, in which the needs and the rights of both women and men are addressed to realise a gender-equal policy outcome (Feenstra & Özerol, 2021). Engendering energy policy resonates with the prevailing academic debate on energy justice and equal access to energy services, as is reflected in international commitments and national policy plans towards a just energy transition.

The narrative of this chapter follows the following outline. After the introduction as given in this section, this chapter describes in Sect. 9.2 the theoretical discourse on existing inequalities and injustices in the energy system that are limiting the implementation of just energy transition policies. We link inequalities with the persisting gendered access to energy as identified in the academic debate of the gender–energy nexus. Injustices in the energy system are described using examples from the energy poverty debate and how energy communities can contribute in mitigating energy poverty. Section 9.3 presents the results of the document analysis of the National Energy Climate Plans (NECPs) of the twenty-seven EU Member States. We identify how the NECPs address energy communities in their ability to address inequalities and injustices in the energy system. In Sect. 9.4, we discuss how energy communities can contribute to a just energy transition that is inclusive and how policy can support the role of prosumers. Section 9.5 concludes the discussion by synthesising the overall chapter content and hints to future research.

9.2 Inequalities and Injustices Limiting a Just Energy Transition

A just energy transition aims to contribute to a just distribution of rights, recognition of needs, and just decision-making within the energy system (Sovacool et al., 2016). The Clean Energy Package strives towards an inclusive and just energy transition for all Europeans in the light of the energy justice debate (European Commission, 2019). This notion of energy justice resonates with the rights-based approach. Rights-based approaches emerged in development studies in the 1990s in response to the failure of development assistance to reduce poverty and inequality (Mukhopadhyay & Meer, 2004; Boesen & Martin, 2007). Although originated in the Global South context, the claims for the right to energy is universal (Tully, 2006). It resonates with the energy justice framework in which recognitional and procedural justice are two main tenets next to distributive justice (Sovacool & Dworkin, 2015). Energy justice was developed in a Global North context in the last decade. Danielsen (2012) developed a gender and rights-based approach applied to energy access as an analytical and strategic framework for researchers and policy-makers. This approach recognises access to energy as a right and acknowledges the institutional barriers women face in realising energy rights due to unequal gender relations. Mukhopadhyay and Meer (2004) position this approach in the engendering policy discourse of gender mainstreaming, providing a decision-making tool for policy-makers to include gender in energy policy design.

Rights-based approaches distinguish right-holders from duty-bearers (Mukhopadhyay & Meer, 2004; Boesen & Martin, 2007; Danielsen, 2012). The state, international community, civil society and the market are identified as duty-bearers. They have the obligations to realise access to energy services for all. Energy communities can be seen in this perspective as duty-bearers, providing environmental, economic or social community benefits to its members or shareholders. Beneficiaries of energy services, energy consumers and the members or shareholders of RECs, are the right-holders. The rights-based approach is used to analyse how gender and other social relations influence the realisation of people's right to access energy. Institutional barriers to women's access to energy are considered rights failures. These failures concern the lack of recognition of women's energy needs, women's energy knowledge and their contribution (i.e., recognition failures), resulting in the unequal distribution of control over energy resources and benefits between women and men (i.e., redistribution failures). Different rights failures can be translated into rights claims at different institutional levels (Danielsen, 2012). Additionally, rights might be difficult to claim due to a lack of knowledge about these rights with the institutions that need to enforce them.

Rights-based approaches focus on achieving basic entitlement to energy services for all. Gender analysis reveals the unequal power relations amongst disadvantaged people and between social groups in society that persist in the EU despite the aim of social inclusion in the Clean Energy for All package of the European Union (Feenstra & Özerol, 2021, Feenstra & Clancy, 2020). The gender and rights-based framework focuses on rights failures, i.e., identifying barriers to realising rights through governance mechanisms at different levels. Danielsen (2012) analysed four decades of gender and energy in development using the gender and rights-based framework. She identified persistent rights failures in the energy system over forty years (1970–2010). Table 9.1

Table 9.1 Persistent rights failures in the energy system through a gender lens

Policy issues in the gender–energy nexus	*Rights failures emerging over time*				
	1970s	1980s	1990s	2000s	2010s
The crisis of biomass degradation	*Lack of recognition of unequal gender relations in the energy system *Gender inequitable access to and control over resources and benefits from energy related development interventions *Lack of recognition of women's knowledge in energy management *Gender inequitable decision-making at all levels in the energy system and the exclusion of women from energy related decisions that affect their lives				
The crisis of women's time		*Lack of recognition of the economic value of women's work making their labour contribution invisible in the energy system at all levels *Lack of addressing women's total energy needs for reproductive and productive purposes			
Poverty alleviation			*Gender disparity in ownership of productive assets required to access and control energy services *Insufficient legal and regulatory frameworks to promote gender equitable access to energy through the market		
Climate change				*Gender inequitable access to financial services resulting in unequal access to and benefits from renewable energy	
Urbanisation					*Gender inequitable access to modern energy in poor urban areas

Source: Adapted from Danielsen (2012)

provides a comprehensive overview of forty years of energy and development work.

Although energy needs are different for each individual, many energy projects, including RECs (Guyet et al., 2021), target users as homogeneous entities. The historical overview presented in Table 9.1 demonstrates that gender and energy issues need to be addressed as a public policy problem to overcome persistent rights failures. Danielsen (2012) stipulates that from a gender and rights perspective, the persistency of rights failures in the energy system represents governance malfunctions in general and accountability malfunctions specifically, in particular at the national and international levels. When taking a right-based approach, the scope moves beyond the individual and makes access to energy a policy concern. To overcome replication of gender rights failures, energy policy should reflect on the gendered failures in the right to energy.

Applying a gender-rights-based approach in energy policy can change this pattern of rights failures. According to Danielsen (2012), gender-rights-based energy policy has a transformative perspective in empowering women and transforming gender relations to be more equal. The underlying theory of change is that in order to create more equitable energy governance structures and institutions, the rights content of energy policy needs to be strengthened. Citizens have rights and responsibilities in the energy system that moves beyond the unequal power relations between individuals. RECs resonate with this theory of change by empowering their members to become owners of their own energy production and create more equity in the energy system for citizens to claim their right to affordable and clean energy (Standal et al., 2018). A rights-based approach leads the way to analyse energy access from a social justice perspective. In the 2010s, the concept of energy justice has emerged in energy policy research to analyse the ethical dimension of the distribution of costs and benefits of the energy transition to society and more particularly to actors involved in the energy system (Miller et al., 2015; Sovacool et al., 2016). Energy justice is developed as a conceptual, analytical and decision-making framework around three tenets: distribution, recognition and procedural (Sovacool & Dworkin, 2015). The rights-based approach is the feeding ground for applying the energy justice framework in gender analysis of energy policy.

The quest for a just energy transition is rooted in the energy justice debate. Energy justice is an emerging framework in energy social science developed to analyse energy transition and energy policy. The energy justice concept aims to contribute to a just energy transition with a just distribution of rights, recognition of need, and just decision-making within energy systems (Sovacool et al., 2016). Energy transitions of different countries take place within a specific governance context, which has implications for energy justice. The key policy goal of the energy transition is decarbonisation: moving from fossil fuels to renewable energy sources and increasing energy efficiency, an aim that was agreed collectively by EU countries and beyond. The process of decarbonisation requires policy-makers to pay close attention to distributive justice: to prevent the unequal distribution of risks, cost-benefits and rights; a key aspect needing attention is monitoring and measuring energy poverty.

Energy poverty is understood as the inability of citizens to afford the energy consumption needed for a decent and comfortable life (Bouzarovski & Petrova, 2015). Conceptually, energy poverty is rooted in the British term 'fuel poverty', established by Brenda Boardman (1991). Boardman recognised that cold homes were having detrimental impacts on the lives of their inhabitants (especially health impacts), and that these impacts were caused by a combination of low income, high energy bills and energy inefficiency in households. The political agenda stemming from this has spread to other European countries and beyond, relabelled 'energy poverty' in the process. Between 2009 and 2019, several legislations designed to address energy poverty were enacted by the European Parliament. Directive 2009/72/EC on common rules for the internal market in electricity was the first to introduce a mention to energy poverty as a growing problem in the EU and to urge Member States to put in place national criteria for the definition of vulnerable consumers and to develop national action plans aimed at decreasing the number of people suffering situations of energy poverty. More recently, in the light of the Clean Energy Plan, Regulation (EU) 2018/1999 and Directive 2019/944 supported the role of RECs to provide affordable clean energy for all Europeans and they have called for energy poverty criteria to be defined and policies developed by national governments to alleviate energy poverty, mentioning RECs as actors in this policy area. EU

Members are obliged to report its incidence at Member State level through indicators, along with a mitigation target and policies when a significant number of energy vulnerable households is identified, starting from the premise that any proportion of households in energy poverty can be considered to be significant.

Energy poverty has a strong gender dimension since more women than men are living in poverty. Feenstra and Clancy (2020) identified three gendered dimensions of energy poverty:

1. Economic: e.g., women with low incomes are disproportionately found as heads of households either as single-parent families or, due to their greater longevity than men, living alone at pensionable age.
2. Biological/physiological: e.g., age is a significant factor in dealing with heat and cold stress, with young children and older people being particularly vulnerable. Women are also considered to be more sensitive to ambient temperature than men.
3. Socio-cultural: women's energy needs and consumption patterns differ when compared to men, but also among women, factors like marital status and employment influence energy consumption.

The complexity of energy poverty and the social dimension of the phenomenon is gaining recognition among energy actors. RECs are increasingly aware of energy poverty and gender inequality in access to sustainable and affordable energy services. However, barriers are recorded by energy RECs to address energy poverty and to become more inclusive (Guyet et al., 2021). The lack of an enabling policy framework is mentioned as one of them. The European Commission continues to develop new legislation in which a just energy transition is enforced, and EU funding is available for implementation and decarbonisation projects. But the national governments of the EU Member States need to implement and enforce this EU legislation in national action plans supporting the local energy initiatives, like RECs. In the following section, we analyse to what extent the NECPs of the EU Member States recognise the energy rights of vulnerable energy users and whether the EU Member States develop a supporting policy for RECs to address energy poverty and gender inequality in the energy system.

9.3 NECPs: Recognising Energy Rights

The EU Governance Regulation on Energy and Climate Action Directive requires Member States to submit NECPs to support the EU's climate action and energy transition objectives (Regulation (EU) 2018/1999). NECPs need to convey a State's national objectives, targets, and contributions to the Energy Union's various dimensions, such as decarbonisation, energy efficiency, energy security and the internal energy market. Specifically, Member States must assess how many households live in energy poverty, taking into account the necessary domestic energy services needed to guarantee basic standards of living in the relevant national context and report on this, as well as policy incentives to mitigate energy poverty. These NECP documents submitted by the national governments of the EU Member States reflect the policy choices of those nations towards a just energy transition. The NECPs serve as a reflection of the policy discourse at the national level. The NECPs are the unit of analysis for our analysis.

The EU legislator requires all European Member States to implement EU directives into national legislation. The NECPs entail a description of how the respective national legislator plans to transpose the EU legislation. In this light, the current NECPs must entail the transposition of the recast of the renewable energy directive (REDII) into national legislation and with that Article 22 on RECs. In the context of this chapter, paragraph 4 is of particular interest, which dictates the national legislator to "provide an enabling framework to promote and facilitate the development of renewable energy communities" (REDII, para. 4). Such an enabling framework removes unjustified regulatory and administrative barriers for RECs. It also ensures the participation of "all consumers, including those in low-income or vulnerable households" (REDII, para. 4 (f)). In addition, REDII provides "opportunities for renewable energy communities to advance energy efficiency at household level and helps fight energy poverty through reduced consumption and lower supply tariffs" (REDII, Recital 67). These requirements entail a drastic shift in approaching vulnerable and energy-poor households. Participation in RECs usually means becoming a member and, with that, a co-owner of the community. However, low-income households are usually not in the position to (co-)own anything. In most countries, e.g., in Germany, the bottom five deciles own 1.4 per cent of net wealth (the lowest is with -1.2 per cent in debt) while the top four deciles own 94.8 per cent of net wealth (Grabka & Halbmeier, 2019). The ongoing COVID pandemic is likely to widen this gap. Economic inequality is equally high in other European countries (Piketty, 2016). Without savings and on an income that hardly covers daily expenses (Eurostat, 2018), wealth accumulation and (co-)ownership are hardly feasible for such groups in society.

With the requirement to enable low-income and vulnerable households to participate in RECs, the EU legislator addresses the structural root causes for such households' inability to become co-owners and accumulate wealth. This is a first step towards bringing together issues of social, economic and energy policy within the just transition framework. This approach has a possible effect far beyond sustainable energy transitions. It requires national legislators to enable formerly economically disengaged passive households that are forced into economic lethargy by low wages or unemployment (a phenomenon particularly common among young people in Southern Europe) to become active energy transition co-owners with access to its benefits. Vulnerable households become agents instead of energy subsidy recipients to cushion transition costs. Economic participation in a discriminating capitalist market economy, formerly exclusively beneficial for those with some form of capital, be that economic, social or cultural, opens up for the discriminated working poor and unemployed.

From a gender perspective, these dynamics are of particular importance: in 2015, women were more likely to experience poverty or social exclusion than men by 1.4 percentage points (the rate for women was 24.4 per cent, while for men it was 23.0 per cent). Almost 50 per cent of all single parents were at risk of poverty or some form of social exclusion. Women are particularly affected as they make up almost 85 per cent of all one-parent families in the EU (Malgesini et al., 2017). Again, these are dynamics likely worsened by the COVID pandemic. At the same time, the little research available on women's economic and energy practices points out particular needs of women regarding their engagement in energy transitions and with that in participating in RECs. The same applies to vulnerable consumers (Hanke & Lowitzsch, 2020). For an enabling framework to be effective, it must find new ways of engaging with these groups in serving their needs. In the end, such a framework provides agency to individuals by enabling their participation in a collective.

So far, the European legislator enables general benefits of RECs but refrains from providing details on the design of an enabling framework to achieve inclusive participation in RECs. It therefore lies within the responsibility and capacity of national legislators to provide a policy mix fit to achieve the above-discussed empowerment of low-income and vulnerable households. In this light, the NECPs should outline national legislators' concrete policy steps towards this goal. At this point, we acknowledge the limitation of a NECP analysis. The NECP documents often do not include all national legislation on energy communities, as Germany and the Netherlands demonstrate. Therefore, the analysis of the NECP documents does not automatically entail all details relevant to this contribution. In addition, many NECPs refer to other pieces of national legislation, which is not part of the analysis of the present study. The analysis, however, illustrates the awareness of the policy-makers on the raised question of injustices and inequalities in the energy transition of their country and the role of RECs to contribute to a just energy transition.

In this light, a content analysis was conducted to analyse the final NECP documents (as of April 2021 the English version) of all twenty-seven European Member States with respect to the question of whether the national legislator provides an enabling framework for (renewable) energy communities to empower vulnerable households and women to participate in the energy transition. The analysis was extended to the question of whether (renewable) energy communities are linked to mitigating energy poverty. In doing so, the following keywords were used in a lexical search: "(renewable) energy community, energy poverty, enabling framework, poor, low-income, vulnerable, empowerment, gender, women". In the next step, text segments were identified where at least two of the keywords overlap. Based on this work, the following codes were established: "REC mentioned, CEC mentioned, enabling framework for REC/CEC, social inclusion in REC/CEC, energy poverty mitigation, REC/CEC and energy poverty mitigation, gender".

As a result, the content analysis of the NECPs reveals that a majority of Member States do not provide a distinct enabling framework for RECs in their NECPs. Out of twenty-seven only six, namely Bulgaria, the Czech Republic, Greece, Italy, Portugal and Spain, mention the importance of enabling the participation of vulnerable households in RECs, i.e., as a means to fight against energy poverty. Currently, only Italy, Portugal and Spain provide more detailed policy measures to achieve this aim in their NECPs. Greece includes in its definition of a REC the social purpose it must fulfil. Tackling energy poverty is explicitly mentioned as one of its main purposes. The remaining NECPs do not establish this linkage yet, at least not in their NECP. None of the NECPs mentions the specific needs of women when empowering their participation in RECs. Some NECPs mention the development of additional legislation to support RECs further or refer to existing legislation. In the analysis, it was not possible to include additional legislation in the approach that was used. Therefore, it is possible that also other countries have additional legislation in place that supports the participation of all consumers and especially of vulnerable groups through RECs, but this national legislation and policy is not reported in NECPs.

9.4 Discussion

Despite the political attention for just transitions that benefit all, there is a policy gap in relation to gender in the low-carbon energy transition. On the policy level in the EU, there has been an emerging focus on engaging consumers to foster sustainable energy consumption and to empower people to become managers of their energy needs (Standal et al., 2018). However, the dominated economic perspective observed in the NECPs focusing on efficiency in the energy transition and decarbonisation tends to ignore social aspects. Energy policies thus do not consider that women and men have different opportunities and ways of engaging with energy in policy, research and business (Standal et al., 2018; Pearl-Martinez & Stephens, 2016; Clancy & Röhr, 2003). In order to establish a better understanding of energy users' situations, more awareness of national legislators for the particular needs and lived experiences of these groups is necessary.

In the search for a just-energy transition, the question emerges how to design an energy policy that reflects the rights and needs of all energy users. Energy technology innovations, such as renewable energy sources and reducing energy consumption, are core aspects of this aim of universal access to sustainable energy. Increasing awareness on the lack of access to sustainable and affordable energy services in the European Union is reflected in the Right to Clean Energy for all Europeans in the EU Electricity Directive 2019/944 of 14 June 2019. The rights and needs of vulnerable energy consumers are gaining attention from policy-makers at all levels within the EU (Feenstra & Clancy, 2020). National governments should acknowledge the legally binding character of the EU directives stating that inclusive participation is a legally enforceable right of under-represented groups.

On the policy level, there has been an emerging focus on engaging consumers to foster sustainable energy consumption and to empower people to become managers of their energy needs (Standal et al., 2019). Women are the primary users and managers of energy at home. Therefore, Makan (1995) argues for an energy policy based on an interactive process directly informed by women and men and their respective needs, rights, interests and knowledge. In the current energy policy design, there is little room for interactive processes that can reflect users' needs, rights, and interests (Feenstra, 2002; Clancy et al., 2017). Energy policy formulation seems to take place entirely at the national level and in a top-down manner. This is partly explained by the supply-oriented focus of energy policy until the 1990s. In the aftermath of the global oil crisis of 1973 and the 1979 energy crisis, security of energy supply was the main focus of national policy-makers. They did not feel the need to consult users, although citizens might have varying preferences about what energy services to use (Feenstra, 2002). The EU Commission required an interactive stakeholder process in drafting the NECPs. National governments were obliged to consult stakeholders and create an interactive process with actors, like online citizens consultations. They had to report in their NECPs to which extent they had consulted stakeholders. However, the voices of the energy-poor remain unheard, and while a local involvement characterises RECs, their national visibility is limited. A strong fragmentation in national policy creates silos in which social issues, like social inclusion and poverty alleviation, are dealt with by welfare policies (Clancy et al., 2017). As a result, stakeholder participation in designing the NECPs is fragmented and limited by not reflecting the voices of the under-represented in the energy transition and specifically the energy poor.

Energy policy discourses often lack a clear notion of who the energy consumers are, and as a result, they fail to recognise gender-specific needs. Analyses of energy policies through a gender lens indicate a gender gap in energy access (Clancy et al., 2017; Clancy & Feenstra, 2019; Feenstra & Clancy, 2020). Women are over-represented among the 'energy poor', those energy consumers that lack sufficient access to energy sources to use the energy they need for heating, cooling, cooking and using appliances. This is mainly caused by the income gap between women and men, women being over-represented in single-parent households and the demographic gender gap that puts more women at risk of poverty than men and for a longer lifetime (EIGE, 2017). More research on the energy practices of these groups and especially the lived experiences of the vulnerable energy users in energy poverty could contribute towards a better understanding of existing injustices and inequalities in energy access (Middlemiss et al., 2020). More gender-disaggregated data on participation in RECs and insights into what limits RECs to become more inclusive in their members and activities could inform policy-makers to design enabling frameworks.

The energy transition is often presented as being dependent on technological innovation and market mechanisms. Replacing fossil fuels like coal with RES in electricity productions is highly beneficial and has been hailed as a win-win solution that reduces emissions and ensures continued energy security. Men are still participating more in local energy initiatives and are over-represented as prosumers (Standal, 2018; Standal et al. 2018). In Norway and the UK, prosumers (who have invested in rooftop household solar systems) are also generally perceived and communicated in media and advertisement as 'techno-savvy, middle-class men with high environmentalist interests' (Standal et al., 2019). Though the decisions to become prosumers were made jointly, there was a gendered division of labour where men took care of the process of becoming prosumers, while women had a notable disinterest in the technical side, but took responsibility to change their energy practices such as doing laundry when the sun was shining (Standal et al., 2019).

Our hypothesis is that in order to be just, the current energy transition has to be inclusive and 'leave no one behind', as expressed in the Green Deal of the European Union (European Commission, 2019). Women's role is invisible in the current energy transition in the EU. The number of women working with RES is higher than for the traditional energy sector (IRENA, 2019). However, jobs held by women are mostly low-paid and non-technical, and there is little prioritisation in the industry towards an inclusive energy workforce (Pearl-Martinez & Stephens, 2016), and there is still a gender gap in Science Technology Engineering and Mathematics (STEM) education recruitment (Clancy & Feenstra, 2019). Combined with the low participation of women as scientists and engineers, it is assumed that women working in the energy sector are represented less in decision-making and more in supporting roles (Clancy & Feenstra, 2019). Research on participation in local energy initiatives in the Netherlands found that women considered themselves to be time-poor and did not want to be involved in committees, and even when they were involved, they did not volunteer for leadership roles (Clancy et al., 2017). On the other side of the energy system, women as energy consumers are equally invisible. Energy poverty reveals the inequalities within society and has a strong gender face. Energy poverty is only an emerging concern in the Netherlands with no gender-disaggregated data and no special attention or policy interventions to mitigate the feminisation of energy poverty (Clancy et al., 2017). However, women in the EU share the same gender inequalities: a gender income gap (partly due to the 'breadwinner' culture), ageing and over-representation as single-parents (EIGE, 2020).

9.5 Conclusions

The research question of this chapter is: what is the potential of energy communities to contribute to a more just and inclusive energy transition and how can European policy support them? To answer this research question, we studied the mutual engagement of female prosumers and energy communities and the potential of energy communities to contribute to a more just and inclusive energy transition. We identified through a discussion of the gender–energy nexus literature that the gender inequality of societies limits the potential of female prosumers to use their agency to participate in energy transition processes. Clancy et al. (2017) demonstrate that since more women live in poverty and occupy energy-inefficient houses as tenants, women face both financial and legal (as tenants) obstacles to invest in energy efficiency. Policy interventions can contribute to overcome these challenges and can stimulate to use the potential of female prosumers to foster just transitions. The EU Green Deal asserts that no one should be left behind in the energy transition process. To recognise women's needs and acknowledge their rights to clean and affordable energy services, energy policies need to overcome their gender-blindness. However, the implementation of such encompassing plans still lags behind, as the present analysis of the NECPs demonstrates.

Even renewable energy communities, which are likely to promote the empowerment of people to claim their right to energy, are dominated by male members. Participating in renewable energy communities (RECs) is a voluntary act that requires time, energy and skills. Although RECs are open to all members with no discrimination, women face barriers to commit themselves because of the other duties within the household and their lack of available time (EIGE, 2020). Another obstacle for women to participate as prosumers in energy communities is their limited investment capacity for membership shares because of the low paid jobs they occupy more often than men (Fraune, 2015). RECs can become more inclusive by creating a socio-ecological contract of enabling policy measures of RECs like a differentiation in share tariffs or membership fees and collaboration with local community partners to reach vulnerable energy users (Guyet et al., 2021). National policy should support RECs in their endeavour to become more inclusive by establishing enabling frameworks like financial support and supportive legislation that will contribute to the empowerment of RECs as inclusive actors in the energy system.

The work of Winther et al. (2020) in the Global South illustrates how renewable energy programmes can be shaped to empower women and transform gender relations at household and community levels. It is an interesting echo of the current transformation in the Global North, especially in the light of the feminisation of energy poverty due to the strong gender face of poverty, the income gap between men and women in the EU, the increase in single-female-headed households and the demographic ageing gap putting more elderly women in poverty than men. In the endeavour to achieve an inclusive and just energy transition, the time is right to include women as stakeholders in energy policy. When the right to energy is universal, the policy should recognise the needs and interests of all energy consumers. This does not need to be a new endeavour, as we can use the lessons learned from energy transition in the Global South to stimulate inclusive RECs and design enabling policy frameworks towards a just energy transition.

References

Bell, S., Judson, E., Bulkeley, H., Powells, G., Capova, K., & Lynch, D. (2015). Sociality and Electricity in the United Kingdom: The Influence of Household Dynamics on Everyday Consumption. *Energy Research and Social Science, 9*, 98–106.

Boardman, B. (1991). *Fuel Poverty: From Cold Homes to Affordable Warmth*. Belhaven Press.

Boesen, J., & Martin, T. (2007). *Applying a Rights-Based Approach – An Inspirational Guide for Civil Society*. Danish Institute for Human Rights.

Bouzarovski, S., & Petrova, S. (2015). A Global Perspective on Domestic Energy Deprivation: Overcoming the Energy Poverty–Fuel Poverty Binary. *Energy Research & Social Science, 10*(0), 31–40.

Clancy, J., Daskalova, V., Feenstra, M., Franceschelli, N., & Sanz, M. (2017). *Gender Perspective on Access to Energy in the EU*. Study for the FEMM Committee of the EU Parliament. PE 596.816.

Clancy, J., & Feenstra, M. (2019). *Women, Gender Equality and the Energy Transition in the EU*. FEMM Committee of the EU Parliament. http://www.europarl.europa.eu/RegData/etudes/STUD/2019/608867/IPOL_STU(2019)608867_EN.pdf

Clancy, J. S., & Röhr, U. (2003). Gender and Energy: Is There a Northern Perspective? *Energy for Sustainable Development, 7*(3), 44–50.

Danielsen, K. (2012). *Gender Equality, Women's Rights and Access to Energy Services: An Inspiration Paper in the Run-Up to Rio+20*. Ministry of Foreign Affairs of Denmark.

EIGE. (2017). *Gender Equality Index 2017 - Measuring gender equality in the European Union 2005–2015*. http://eige.europa.eu/gender-equality-index. Accessed 28 April 2021.

EIGE. (2020). *Gender Equality Index 2020: Digitalisation and the Future of Work*. http://eige.europa.eu/gender-equality-index. Accessed 28 Apr 2021.

European Commission. (2019). *Clean Energy for All European*. Publication office of the European Union.

European Parliament and European Council. (2018). Directive (EU) 2018/2001 of the European Parliament and of the Council of 11 December 2018 on the Promotion of the Use of Energy from Renewable Sources (Recast).

Eurostat. (2018). *In-Work Poverty in the EU*. European Commission. https://ec.europa.eu/eurostat/web/products-eurostat-news/-/DDN-20180316-1

Feenstra, M. (2002). *Towards a Gender-Aware Energy Policy: A Case Study from South Africa and Uganda*. Master thesis, University of Twente.

Feenstra, M., & Clancy, J. (2020). A View from the North: Gender and Energy Poverty in the European Union. In J. Clancy et al. (Eds.), *Engendering the Energy Transition*. Palgrave Macmillan.

Feenstra, M., & Özerol, G. (2021). Energy Justice as a Search Light for Gender-Energy Nexus: Towards a Conceptual Framework. *Renewable and Sustainable Energy Reviews, 138*(110668), 1–11.

Fraune, C. (2015). Gender Matters: Women, Renewable Energy, and Citizen Participation in Germany. *Energy Research and Social Science, 7*, 55–65.

Fuller, S. (2019). The Politics of Energy Justice. In K. Hancock & J. Allison (Eds.), *Oxford Handbook of Energy Politics* (pp. 1–19). Oxford University Press.

Grabka, M., & Halbmeier, C. (2019). *Vermögensungleichheit in Deutschland Bleibt Trotz Deutlich Steigender Nettovermögen Anhaltend Hoch*. DIW Wochenbericht. https://doi.org/10/gjvfs6

Grübler, A., Wilson, C., & Nemet, G. F. (2016). Apples, Oranges and Consistent Comparisons of the Temporal Dynamic of Energy Transitions. *Energy Research & Social Science, 22*.

Guyet, R., Hanke, F., & Feenstra, M. (2021). *Energy Communities and Energy Poverty: Moving Towards a New Social and Ecological Contract* (ENGAGER Policy Brief Working Group 4, No. 3). http://www.engager-energy.net/wp-content/uploads/2021/03/WG4-policy-brief-March-31.pdf

Hanke, F., & Lowitzsch, J. (2020). Empowering Vulnerable Consumers to Join Renewable Energy Communities – Towards an Inclusive Design of the Clean Energy Package. *Energies, 13*(7), 1615, 1–27.

IRENA. (2019). *Renewable Energy: A Gender Perspective*. Abu Dhabi: International Renewable Energy Agency (IRENA).

Listo, R. (2018). Gender Myths in Energy Poverty Literature: A Critical Discourse Analysis. *Energy Research and Social Science, 38*, 9–18.

Makan, A. (1995). Power for Women and Men; Towards a Gendered Approach to Domestic Energy Policy and Planning in South Africa. *Third World Policy Review, 17*(2), 183–198.

Malgesini, G., Cesarini-Sforza, L., Babović, M., Sverrisdóttir, M., & Mareková, S. (2017). Gender and Poverty in Europe. *Women and Poverty, 24*.

Middlemiss, L., Straver, K., Hesselman, M., Tirado Herrero, S., Feenstra, M., Hargreaves, T., Meyer, S., O'Sullivan, K., Sareen, S. & H. Thompson. (2020). *Making the Most of Qualitative Evidence for Energy Poverty Mitigation: A Research Agenda and Call for Action*. ENGAGER Policy Brief. No. 3.

Miller, C., Richter, J., & O'Lary, J. (2015). Socio-Energy Systems Design: A policy Framework for Energy Transitions. *Energy Research and Social Science, 6*, 29–40.

Mukhopadhyay, M., & Meer, S. (2004). *Creating Voice and Carving Space: Defining Good Governance from a Gender Perspective*. KIT Royal Tropical Institute, The Netherlands.

Pearl-Martinez, R., & Stephens, J. (2016). Toward a Gender Diverse Workforce in the Renewable Energy Transition. *Sustainability: Science, Practice and Policy, 12*(1), 8–15.

Piketty, T. (2016). *Das Kapital im 21. Jahrhundert*. Translated by Ilse Utz and Stefan Lorenzer. 1. Auflage in C.H. Beck Paperback 6236.

Sovacool, B., Burke, M., Baker, L., Kumar Kotikalapudi, C., & Wolkos, H. (2017). New Frontiers and Conceptual Frameworks for Energy Justice. *Energy Policy, 105*, 677–691.

Sovacool, B., & Dworkin, M. (2015). Energy Justice: Conceptual Insights and Practical Applications. *Applied Energy, 142*, 435–444.

Sovacool, B., Heffron, R., McCauley, D., & Goldthau, A. (2016). Energy Decisions Reframed as Justice and Ethical Concerns. *Nature Energy, 1*(5), 16024.

Standal, K. (2018). *Challenges of Gender, Power and Change in Solar Energy Interventions in Rural India: Imagined Beneficiaries and the Makings of Women's Empowerment in the Village Electrification Project*. PhD. Dissertation, University of Oslo.

Standal, K., Talevi, M., & Westskog, H. (2019). Engaging Men and Women in Energy Production in Norway and United Kingdom: The Significance of Social Practices and Gender Relations. *Energy Research and Social Science, 60*(101338), 1–9.

Standal, K., Winther, T., & Danielsen, K. (2018). Energy Politics and Gender. In K. Hancock & J. Allison (Eds.), *Oxford Handbook of Energy Politics* (pp. 1–21). Oxford University Press.

Tully, S. (2006). The Human Right to Access Electricity. *The Electricity Journal, 19*(3), 30–39.

Winther, T., Ulsrud, K., Matinga, M., Govindan, M., Gill, B., Saini, A., Brahmachari, D., Palit, D., & Murali, R. (2020). In the Light of What We Cannot See: Exploring the Interconnections Between Gender and Electricity Access. *Energy Research and Social Science, 60*(101334), 1–18.

10

Housing Communities as Low-Carbon Energy Pioneers; Experiences from The Netherlands

Lidewij Tummers

10.1 Introduction

Co-housing is an umbrella term for a wide variety of housing clusters with a high degree of residents' involvement in the design and management decisions. In attempts to structure research on its manifold forms of projects, many scholars have proposed definitions of what co-housing is (Lang & Stoeger, 2017; Giorgi, 2020; Locatelli et al., 2011; Vestbro, 2000 and 2010). Co-housing is a highly decentralised form of housing, following the observation of Ache and Fedrowitz (2012) that "in co-housing the organisational entity overlaps with the spatial entity". This is mostly paired with high ambitions in making housing affordable,

L. Tummers (✉)
Fakultät für Architektur und Landschaft, Institut für Umweltplanung,
Leibniz University Hannover, Hannover, Germany
e-mail: tummers@umwelt.uni-hannover.de

© The Author(s), under exclusive license to Springer Nature Switzerland AG 2021

F. H. J. M. Coenen, T. Hoppe (eds.), *Renewable Energy Communities and the Low Carbon Energy Transition in Europe*, https://doi.org/10.1007/978-3-030-84440-0_10

convivial and sustainable.[1] Furthermore, international literature reviews and fieldwork showed that resident-led housing projects are pioneers in energy transition (Tummers, 2015a, 2017).

In this chapter, based on case studies, I will argue that co-housing can also be seen as an energy community in the EU renewable energies directive definition as self-organised groups of households around an energy-related activity. This relies on two pillars of co-housing: first, the general idea that residents form an intentional community (Jarvis, 2015; Sargisson, 2012), and second the territory of this community in the form of 'housing'. Housing is currently an important component of our energy consumption, and increasingly a location for energy production. Although co-housing organisations do not solely aim to generate and distribute (renewable) energy, many if not most of the decisions they take are energy-related. Hence, addressing co-housing associations and incorporating them into the implementation of the Clean Energy package (see Chap. 2 by Roberts in this volume) will contribute to achieving the Green Deal/climate change objectives.

While housing clusters that are captured by the umbrella term 'co-housing' share the main characteristics of self-governance, with the aim to create affordable, convivial and sustainable dwellings, a more detailed portrait of co-housing in the Netherlands shows there exist many varieties of co-housing (see Table 10.1). The first question this chapter aims to answer is thus: 'which of the manifold forms of co-housing typologies potentially qualify as (renewable) energy communities according to the definition of the Clean Energy package?' The second question concerns the relevance of relating energy- and housing communities, asking: 'does co-housing make a stronger contribution to the energy transition than mainstream/standard housing?' The answers suggest that co-housing as intentional communities can lead to potentially higher energy performance of housing, and provides instrumental functions for energy transition. The chapter concludes by summarising how these features can benefit from an approach as an energy community, while contributing to the objectives of the Clean Energy Package.

[1] The bottom-up self-governance of co-housing projects differs from gated communities that also tend to be defensive, expensive and exclusive.

Table 10.1 Varieties of Dutch resident-led housing (author, August 2014) Updated 2021

Dutch terminology	Translation/definition	Period	Tenure	Characteristics
(A) Centraal wonen	Co-housing following the international co-housing model: see www.lvcw.nl	First generation 1980s	Social rental in partnership with housing association (see: *zelfbeheer*)	Community-building is the central factor. 6–8 households share a kitchen and other rooms for everyday use; these groups cluster into larger projects with common facilities and management.
(B) Zelfbeheer	Self-management:	Often former squats legalised in the 1990s	Residents do not own the premises but form an association in order to maintain and manage it	Typically in large cities, and in other regions for which WBVG is a partner. Mostly renovated or re-used complexes/ heritage buildings.
(C) MW^2: Mens- en Milieu-vriendelijk Wonen en werken	Human and environmentally-friendly living and working	1990s	Mixed tenure: owner-occupied and rental	Typically in medium-size towns. Newly built individual houses around a common garden with shared spaces and high environmental ambitions.
(D) Woongroep voor ouderen	Community for seniors	Widespread since the late 1980s	Individual units within standard rental procedures with shared space and facilities rented from residents' association (RA)	Collectively managed common rooms and without structural institutional interference after the building phase, new tenants chosen by RA.

(*continued*)

Table 10.1 (continued)

Dutch terminology	Translation/definition	Period	Tenure	Characteristics
(E) Eco-dorp	<u>Eco-village</u>	Movement since the 1980s	Obtaining land through the availability of brownfield sites such as former airports or institutions	Larger scale initiatives that aim for holistic renewal: energy-transition, food-production and so on. Has regained momentum through the CPO-policies.
(F) Collectief Particulier Opdrachtgeverschap (CPO)	<u>Collective Private Client</u> (CPO), equivalent of *Baugruppe* (building groups)	Since 2000 law on private housing development	Predominantly individual home-ownership, or mixed tenure	Often during the design stage common building parts, parking, playground or such are decided and remain in co-management after building. Also called (in Dutch): *Bouwen in eigen beheer* or *Eigenbouw*

10.2 Energy and Co-Housing Research

This chapter results from a PhD research project,[2] which had as its main purpose a search for innovations in resident-led housing that could be relevant for making the wider housing stock more sustainable. This involved analyses of the energy-systems of co-housing, based on multi-case-study analyses. The cases were selected looking at the projects' dimensions, typology, population and locations as representative for different generations of housing and environmental regimes in the Netherlands. For the state of the art of energy-efficient housing in the Netherlands, national policy documents and programme evaluation reports were analysed. This inventory led to four periods of energy and housing typology, each introduced by a rupture with previous policies or regulations (see Table 10.2). From each period, representative co-housing projects were selected with institutional state-of-the-art housing clusters as benchmarks. The cases are listed in Table 10.3. Single cases with unique features (such as the conversion of monumental buildings) were thus put aside.

The fieldwork focussed on a general understanding of the energy-systems chosen and operated by the residents' associations. Information on the initiatives was gathered through 'participative research': during project visits and interviews a dialogue took place between researcher and interviewees, about the motives that led to the choice for one construction or engineering type over other possibilities. The roles of the interviewees varied from residents running the association to consultant, evaluator, partner or professional involved in the project design, construction or management. Energy-related information of projects was found in the technical briefings for the building license, contracting and technical maintenance. Planning documents further provided information into the extent to which energy optimisation had influenced the urban layout and cluster design. During visits, the technical data were discussed and verified with the residents, who often have detailed

[2] An earlier version was published with Prof Dr Andy van den Dobbelsteen (2017) my PhD report; http://abe.tudelft.nl/index.php/faculty-architecture/article/view/1858

Table 10.2 Significant environmental policy periods and pilot projects experimenting with sustainable building

Table 10.2	Year	Context environmental & energy policy	Energy standard in housing	Top-down experimental projects	Self-managed co-housing & energy projects
1985–1995 INTRODUCTION	1989	First environmental policies & standards	Double glazing & some insulation following oil crisis	Ecolonia (1988–1992, Alphen a/d Rijn)	Romolenpolder Haarlem (1989)
1996–2003 INSTRUMENTALISATION	1996	Intro EPC	EPC≤1.6		Groene Marke Zutphen (1993–97)
	1999–2012	Nationaal Pakket Duurzame woningbouw	1998 EP 1.6>1.4		
	1996–1999	SEV stimulation & demonstration programme		Oikos (1997–2005, Enschede)	Increase in number of initiatives
	2000	National Self-building policy	EPC 1.4>1.0		
2008–2012 STAGNATION	2008	Intrduction of the E-label & triple convenant	Average label D-E		Meander, Zwolle (2008)
		Financial crisis	Adjustments of feed-in tariffs;	Building sector paralysed	Overall (C)PO share remains at ±15%
	2012	SBR DuBo catalogus (replacing Nationaal Pakket)	EP 1.0>0.4		

2013–present RE-EMERGENCE	2014	National Energy Agreement	Preliminary energy labels		'Postcoderoos' favourable for co-housing
	2014	Warmtewet (Law on heat supply)	District heating	Smart grid pilot	Kleinverbruiker ≤100 kW; levering vergunningvrij max 10 hh
	2013	'Energie-sprong' (Energy-leap) national programme	Lowering of EP standard delayed to recover from crisis	'Stroomversnelling' (renovation rapid) pilot projects	IEWAN, Lent (2015)
	2015	First time limitation on natural gas production; COP targets	New model for EP calculation (NEN 7120)	'Stroomversnelling' stagnates in government procedures	Urgenda wins lawsuit over national government Eco-villages become partners for local and national authorities

Source: Tummers and van den Dobbelsteen (2015)

Table 10.3 Overview of energy-features of discussed projects

Characteristics of studied projects	Units, tenancy	Building tech	Engineering	Energy-performance	Govmtal programme/subsidies
CW romolenpolder (1990, Haarlem)	46 rental social housing regimes	Concrete, lime-blocks (kalkzandsteen), brick, tropical wood framing	Individual VR gas heating	Before EPC. Rc =2.5 (roof/ facade) following social housing standard	NormKostenSysteem=normatised housing subsidies
Groene Marke (1995, Zutphen)	50, rental and home-ownership, mixed prices; commercial units; common house as shared space	Timber structures and concrete floors	Individual VR & HR gas heating, LTS optional for home-owners; collective rainwater capturing	Before EPC. Now assessed label A or B	National Demo programme (SEV)
Meander (2008, Zwolle)	53, mixed tenure, including workshops and meeting place	Innovative building materials, high thermal insulation standard	Individual HR++ ketels	EP=1.2	

IEWAN (2015, Nijmegen)	Clustered multi-storey 24 rental units; common house plus workspaces	Straw-bale self-built Bufferzones/ semi-collective spaces	Collective pellet heating; collective solar, water recycling through local reedbed	Energy-neutral	Subsidies from local authorities

Source: Tummers, 2016, updated (2019)

knowledge about modifications over the years after construction. They also represent the groups' attitude towards energy and technology.

The part of the research presented here has looked at the engineering of utilities and use of renewable energies in resident-led housing projects and compared residents-led and institutionally provided housing, establishing which differences exist with standard 'mainstream' housing models and how these affect the energy performance. The type and quality of utility engineering influences the energy-efficiency of (residential) buildings, for example through reducing exergy (i.e., system losses). Quantitative empirical data on energy usage in self-managed housing clusters were needed to assess the long-term performance and comparing it to benchmarks from mainstream forms of housing. However, so far very little quantitative empirical evidence or systematic results of monitoring have been published and comparable figures are not sufficiently available. Moreover, closer analyses revealed a number of specific obstacles that hamper an empirically grounded understanding of energy performance in self-developed housing clusters. These will be addressed in Sect. 10.5.

10.3 Co-Housing in the Netherlands

Co-housing is (re-)emerging in Europe as a bottom-up movement of groups of households building, occupying and managing a housing cluster (Krokfors, 2012; van der Woude 2012; Tummers, 2015a, 2015b; Benson, 2014). Co-housing projects exist in many varieties, from Baugruppe to housing communities, but generally comprehend 20–60 one-household units with shared spaces such as playgrounds, bicycle sheds, function rooms and launderettes. The example of Tergbregse.nl (Fig. 10.1) can be read as a prototype of Dutch co-housing in the twenty-first century: developed in a collaborative design process with an important role for the residents' association playing a major role; applying modular design system based on collective choices, but allowing individual extras; mixed tenure (owner-occupancy and subsidised rental); shared spaces and facilities. It differs from gated communities also through a degree of permeability from (or openness towards) the

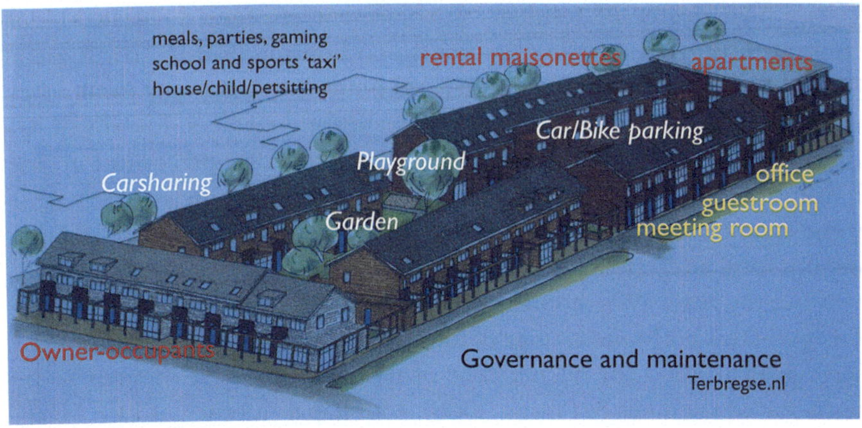

Fig. 10.1 TERBREGSE.NL (Drawing by Hulshof Architekten, 2000, inscriptions by author)

neighbourhood. The inner courtyard is only partly privatised, leaving room for a meeting garden and playground.

Co-housing projects are based on voluntary participation and effectively controlled by its residents. They are characterised by what Ache and Fedrowitz (2012) have called 'an overlap of the organisational and spatial entity' which excludes the larger housing institutions, even coops, that manage a number of locations from a central office, albeit somewhat 'controlled by shareholders or members'. The objectives of co-housing initiatives are primarily to provide safe and healthy living environments; financial profit (housing speculation) has never been signalled as an objective, and is rather a problem disrupting communities.

Some projects are highly value-driven, focussing for example on the inclusion of vulnerable groups or lowering their ecological footprint. Other clusters are more pragmatic in realising that 'naoburskip' ('being good neighbours') and pooling resources offers a range of possibilities that individual households could not afford. Projects from an earlier generation, realised in the 1980s to 1990s, demonstrate that co-housing creates stable and high-quality living environments. The literature classifies them as 'villages in town' (Lietart) 'urban oasis' (Haquebord) or 'pocket neighbourhoods' (Chapin).

The high degree of residents' involvement changes the stakeholder constellation in the housing market. In co-housing, former 'consumers' or 'beneficiaries' of individually serviced homes are becoming joint commissioners and producers of housing, energy and services, forming a so-called 'prosumers' community. This indicates the precondition for housing clusters to classify as 'energy communities', namely that there is some form of organisation and collaboration within the project, and this organisation addresses energy issues.

Since the late 1990s, housing policies have become aware of the implications of this profound change of roles. The Dutch National Government set a goal in 2000 for 30 per cent of new housing development to be resident-led (Remkes & Pronk, 2000). Municipalities such as Den Haag, Rotterdam, Almere and Amsterdam, have adapted land-allocation policies to accommodate self-organised housing development both in urban extensions and inner-city contexts (Roetgerink, 2006). New professions are emerging to accompany groups during the development process or lend support in housing management (Tummers & Palmer, 2019). To date, a stable 15 per cent of housing production appears as resident-led in the Centraal Bureau voor Statistiek (CBS) statistics. It needs to be pointed out however, that these statistics do not include all forms of collaborative housing because they register the formal applicant of building licence. The co-housing research showed that the residents' associations in the Netherlands often depend on collaboration with institutional partners, for example to realise subsidised rental units, obtain land or create micro-grids. The established institutions act as a back-up partner, assuming the formal role of applicant for the building licence while leaving decision-making to a large (but variable) extent to residents' associations. This is reflected in Table 10.1 showing 'hybrid' forms of co-housing.

10.4 Co-Housing and Energy

This section explores the (renewable) energy practices of these co-housing types to answer the question if they qualify as community according to the definition of the EU Clean Energy deal. The cases for this study demonstrate that social practices in self-managed housing clusters include new patterns of consumption and mobility. Initiatives develop energy-related practices that are different from the single standard consumer-unit. The practice of sharing in co-housing thus involves energy-related innovations that advance energy transition in residential clusters, for example, sharing (e-)cars or professional laundry installations. So far, energy-related research into self-developed housing clusters is scarce (Tummers, 2013, 2015a; Palojärvi et al., 2013; Chatterton, 2015). Most studies primarily looked into the benefits of sharing activities, such as cooking and eating meals, laundry installations, play or meeting rooms, or other facilities (see for example Kido & Nakajima, 2011; Stevenson et al., 2013). A Swedish explorative case study calculated that the everyday practices in co-housing reduces CO_2 emissions by 1 ton compared to standard housing (Sundberg, 2014). A Swiss self-assessment found that the housing footprint of Coop residents was less than half that of the average Swiss (Equilibre, 2016).

In addition to social practices, pooling resources makes it possible for co-housing participants to jointly consume and produce renewable energy, and to operate the technical utilities within the project, such as recycling of waste and water, solar or common heating installations. The energy-related 'rationale' of production is far apart from the 'rationale' of usage (see for example: Ingle et al., 2014; Gram-Hanssen, 2013; Subrémon, 2011). Both individuals and groups 'producing' renewable energy need to bridge this gap and harmonise both rationales; for example, in a collaborative housing project investing in PV, participants also need to be concerned with energy distribution—both through technical networks as well as deciding about the range of clients.[3] They may decide to develop a number of inhouse storage options, such as shared e-cars and e-bikes; distribute the electricity following demand, or connect the system to the common spaces exclusively. In this process, they develop innovative solutions for decision-making, maintenance, finance and administration directly related to energy flows and thus become 'prosumers'.

[3] There are limitations on the distribution of 'self-produced energy' amongst others expressed in the so-called '*postcode roos*' defining the delivery area by postal code.

As Gram-Hanssen (2014) has shown, the engineering of utilities mediates between energy flow and social practices. Evaluation of the Dutch co-housing projects has produced insights that can improve the engineering of utilities for self-managed housing clusters. The case studies show that collaborative dynamics offer additional opportunities for low-impact energy at an intermediate scale between the larger network and the individual household. Furthermore, the high degree of self-organisation indicates that they can be seen as energy communities.

Besides technology and social practices, the housing projects are shaped by the institutional environment that imposes building regulations and legal requirements on the design. The landscape of Dutch energy regimes against which the housing initiatives emerge accounts also for the selection of projects from different regimes (ECN 2012). The policy periods relevant to energy can be summarised as follows (see Table 10.2):

- 1985–1995: the introduction of concepts of sustainability and the need for reduction of CO_2 emission and fossil fuel dependency;
- 1996–2006: implementation development and dissemination of instruments such as LCA, Energy Performance Code (EPC), financial incentives, handbooks. European energy directives;
- 2007–2012: stagnation, due to lack of investment capacity and economic priorities of a new coalition in national government;
- 2013–2019: re-emerged sense of urgency, due to societal pressure. A new interest in civil initiatives is changing the relation between public and private initiatives, as is further discussed below.

Table 10.2 shows the landmarks and the selection of cases.

Self-developing housing clusters eliminate the developers' margin (van den Ham & Keers, 2010). Both the fieldwork and literature review showed that rather than reducing the budget, projects invest this margin (Van Voorden et al., 2004). Co-housing initiatives invest especially in state-of-the-art sustainability, and are capable of finding their way in energy policies. For example, they are proportionally over-represented in subsidised innovative housing programmes (see, for example, Van Voorden et al., 2004; van der Reijden et al., 2002). In the second national demonstration programme for sustainable housing 1996–1999 four out of twenty-five housing projects were initiated by residents-collectives and a fifth project follows the co-housing model while a housing association was in charge (reconstructed from Buis, 2000).

In 2014, the national bureau for planning (CPB) and the Netherlands EnvironmentalAssessment Agency (PBL), agents of the government, realised a joined study into the potentials of bottom-up projects. According to their findings, the bottom-up initiatives primarily consist of energy cooperatives, but a large part of the population is not involved in energy transition (CPB/PBL, 2014: 50). The CPB and PBL signal that:

> At macro level, the Netherlands is far behind European countries in the transition towards renewable energies, but this may change with the emergence of bottom-up initiatives. Solar collectors and policies have enabled people with environmental concerns to actively contribute to the solutions. [...] Some energy innovations, such as double glazing, thermal insulation and highly efficient boilers, have reached the 'late majority'. Solar technology has reached the 'early majority'. However, energy-neutral dwellings are still a privilege for a small group of 'early adopters'. (CPB/PBL, 2014: 49, authors' translation)

The reason is thus sought in the motivations of households, assuming the interest in new energy technology increases when the impact becomes easily visible, for example in reduced bills or more comfort. The CPB and PBL advise to connect the technological innovations to local initiatives already engaged in environmental or social concerns.

Dutch co-housing projects, however, demonstrate that such connections have existed since the 1990s. The following section presents some examples to illustrate the choices made for engineering and social organisation in co-housing. These cases also illustrate how buildings are largely shaped by the (affordable) technologies available. This in its turn is influenced by energy policies that emphasise energy savings at one time, or renewable sources at another. Their key-features are summarised in Table 10.3.

10.5 Engineering and the Use of Renewable Energies in Co-Housing

To introduce the principle of sustainable housing, the governmental experiment Ecolonia was built, demonstrating a number of market-available options to reduce energy consumption and use renewable sources (technical briefing Ecolonia 1991). In the same period residents initiated a co-housing model A, Centraal Wonen project. CW Romolenpolder consists of forty-six rental units of different sizes built around a common garden which were delivered in 1992. The project was realised under a subsidised housing regime,[4] which obliged them to involve an established housing association (HA). The project is located in the first eco-district of Haarlem, and the residents proposed a number of sustainable measures, but none of them were accepted by the HA. Interviews with residents and minutes of project design meetings witness the large amount of information the RA gathered to support their argument and thus contributing to the introduction of sustainable building in general. Nonetheless, during the building of Romolenpolder, public housing regimes did not prioritise sustainability.

In 1991, sustainable policies changed focus from building to urban design, applying new principles of pavement, water management and greening in urban expansion areas such as Oikos, Enschede. At the same time, resident associations came increasingly to realise sustainable co-housing project model C. In Zutphen in 1995–6 for example, fifty residential units and workspaces built around a Green Common ('Groene Marke') of about 4000 m^2 were delivered. The residents' association elaborated the design and technical brief with a local architect, who had implemented the then latest insights of sustainability in two earlier projects in the area (technical documentation Grotenbreg Firm, 1992–1997). The association collaborated with a housing association Woningstichting Zutphen (WZHA) to realise a mixed-tenure structure, including 25 per cent low/medium-budget rental units. The WZHA delivered professional services such as financial administration and supervision of the building

[4] Het Normkostensysteem (NKS), Nationale Woningraad, Almere, 1986.

works and acted as formal client for the contractor. The architect office wrote two briefings: one standard (following the WZHA specifications) and a second adding 'sustainability measures'. The contractors had to make two cost-estimates, standard and environmental. The rental units are equipped with the WZHA standard which means small units (social minima) are heated with a standard device (VR, improved efficiency boiler), leaving low-rentals with a relatively larger energy bill, whereas larger units have individual high efficiency (HR, high-efficiency boiler) heating. Home-owners had options to (self-)install wall-heating and solar panels. All roofs were designed for solar energy, with a 30–50° slope and suitable orientation.

Since 1996, the Residents Association Groene Marke captures rainwater and sells it to the inhabitants replacing the use of drinking water for toilet-flushing and washing machines. This also helps to pay for the communal house bills and innovations (interview technical delegates Groene Marke, 2012). Currently they are studying further measures to make the project 'energy-neutral', for example, through the introduction of collective use of geothermals.

SInce 2001, national policies have focused on stimulating self-building rather than intensifying sustainable housing. Several co-housing initiatives benefited from the policy to realise their housing project. The residents' association Meanderhof in Zwolle established a partnership with DeltaWonen, the local housing association (DWHA), and an architect specialised in ecological construction. The project was completed in 2008 to include fifty-three housing units, small businesses, meeting rooms and a communal bicycle shed. The energy concept is based on "high insulation rather than complex technology" (interview residents February 2012), resulting in an average EPC score of 0.65 at a time when the norm was still about twice as high (1.2, see Table 10.1). The project was built from an innovative wood construction with high thermal insulation, using new products such as cellulose, Pavatex panels and insulated wooden window-frames, imported and adapted from a Scandinavian model and applied for the first time in the Netherlands. The flats are however equipped with a conventional heating system (Dutch standard at the time). The experimental building methods required considerable attention during design and build, and an interest for technicalities from the residents. However, in making design choices "we prioritised human interaction, but often this works well together with sustainability, for example in creating common courtyards instead of private gardens" (interview residents February 2012).

The stagnation of the building sector following the credit crunch in 2009, enabled the 'Initiative Ecological Housing Nijmegen' with the acronym of IEWAN to find a building site in the Nijmegen urban extension of Lent. IEWAN created partnerships with Talis (THA), Woningbouwvereniging Gelderland (WBVG, an umbrella organisation for self-managing housing model B), and local and regional authorities. Under a social (affordable) housing regime, twenty-four units were built with common rooms and services that have been inhabited since May 2015. A second, smaller, building provides working spaces and function rooms. Ecological and energy ambitions were high, hence the choice for clustered building in straw-bale construction. The dwellings are situated to capture sun especially in winter, and additional heating is provided with a collective pellet incinerator. Hot water is generated through heat recovery from ventilated air. There is no connection to the natural gas network, a novelty for housing in the Netherlands. Rainwater is captured in a reed bed and recycled for toilet flushing (technical briefings IEWAN and explanation by inhabitants, 2015).

During the building phase, instructive workshops were organised for volunteers who in exchange helped build the outer walls. This added to the challenges for the contractor; in addition to applying innovative technology, the firm was bound to contractual obligations and quality standards of the HA, as well as formally responsible for volunteers on the site. The residents emphasised the learning process that took place between the initiators and their institutional partners, and passed their experience on to other initiatives (interview 2015).

The cases show how co-housing residents engage in decisions that affect energy-efficiency designing and developing their project, including maintenance and renovations. Increasingly, they are also involved in energy supply, for example through common geothermal or PV installations, thus providing other energy services to its shareholders or members. The benefits of generating ones' own energy in the project is that the revenue stays in the organisation to address the needs of the community, such as investing in energy efficiency; and maintain shared spaces such as playgrounds or function rooms. This is a practice shared with many REScoops in the Netherlands (Coenen & Hoppe, 2018; RESCOOP 2016). Finally, as Marckmann, Gramm-Hansson and Haunstrup (2012) found, housing communities are instances for knowledge exchange and have great potential for environmental learning among inhabitants.

10.6 The Potential of Co-Housing as Energy Communities

Some specific energy-related features of co-housing stand out: first, the residents' associations were the driving force towards sustainable building and energy and water systems. Second, when residents are in the lead, the emphasis shifts from technical efficiency to wider criteria, such as flexibility, transparency, hands-on involvement and fair billing systems. Third, the collective action facilitates specific sustainable solutions such as pellet-heating, low-temperature grids, PV systems and water recycling. These solutions are optimised by joined action and learning processes.

These observations bring co-housing close to the EU Clean Energy for All Europeans Package definition of 'renewable energy community' (REC). Residents' associations managing co-housing fulfil the criteria of being a legal entity:

> ...which, in accordance with the applicable national law, is based on open and voluntary participation, is autonomous, and is effectively controlled by shareholders or members that are located in the proximity of the renewable energy projects owned and developed by that legal entity;
>
> The inhabitants are members of a residents' organisation, located on the premises, taking the decisions for joined energy-related investments and practices.
>
> ...has for its primary purpose to provide environmental, economic or social community benefits to its members or shareholders or to the local areas where it operates rather than to generate financial profits;

> The residents' organisation's primary task is to secure affordable, stable and healthy dwelling environments for its members. Any benefits generated will be re-invested in the community through joint decision-making;
>
> and:
>
> …may engage in electricity generation, distribution and supply, consumption, aggregation, storage or energy efficiency services, generation of renewable electricity, charging services for electric vehicles or provide other energy services to its shareholders or members.

The decisions and activities of a residents' organisation are always energy-related in terms of consumption, often concerning distribution of energy systems (shared spaces) and increasingly for production and recycling. Beside direct energy delivery (e.g., photovoltaic or micro-grid), sharing practices involve services to residents such as mobility, function rooms or tools and workspaces, that indirectly involve energy storage or (re)distribution.

Co-housing legal entities (the resident associations) are not always the owner of the real-estate.

Typical for Dutch co-housing is the mixture of tenancy, combining owner-occupancy and rental units (C, E, F), or individual rental contracts with a self-management agreement between owner and occupants (A, B and D). This can be an obstacle, since housing associations are reluctant to investing in energy efficiency because of the so-called 'split incentive' (Gillingham et al., 2012): they do not benefit financially from energy-saving measures they invest in. Nonetheless, the classification of some projects can be diffuse, particularly that types A to E qualify for the criteria, and F is variable depending on the ambitions of the Baugruppe.

The definition practically equals 'energy' with 'electricity' but includes energy efficiency services. In co-housing, charging services for electric vehicles and shared cars are indeed often available or may serve as storage of surplus PV-generation. However, limiting energy to electricity ignores a major potential of co-housing to contribute to energy transition and climate targets.

Initiatives for self-managed housing projects can thus be seen as communities of practice that conceptualise environmental and energetic ambitions in an 'integrated' or 'holistic' way. The selected projects illustrate what this means for the energy system design:

- not only to reduce energy demand, but also to apply clean and renewable resources;
- beside the direct use, also taking into account indirect energy needed for the production of building material, devices and water delivery;
- 'low impact' is not only a standard for design and construction, but also for the use and management of the buildings;
- the social pillar of the sustainability triangle, such as social inclusion and interaction, are high on the agenda.

10.7 Why Is It Important to Include Co-housing as Energy-Communities?

Collectively developed housing projects and the clustering of units open up a range of possibilities beyond individual dwellings. Tillie et al. (2009), for example, point at the possibility to re-use energy flows on an intermediate scale. Clustered development also enables the application of alternative sources, which normally are not affordable for single dwellings; for example geothermal or cogeneration (WKK). Data from demonstration projects indicate further savings may be achieved in the material and piping for collective, high-tech energy utilities. On the other hand, the internal distribution system expands, which leads to transport losses for heat or hot (tap) water distribution and inefficient heat-recovery systems from ventilated air (Stofberg et al., 2000: 115). It is difficult to predict in which way the balance tilts, because until now research has predominantly focused on reduction of energy used for heating in individual housing units.

An example is the presence of shared spaces such as the glazed corridors in Romolen: widened corridors, halls, covered courtyards and enlarged staircases (Williams, 2005; Fromm, 1991). Such intermediate spaces are important for informal contacts, while they also form buffer zones between the outdoor temperature and indoor heated dwellings, which IEWAN is making use of. This combination of space to enhance informal meetings between residents, and buffers to save energy, influence the engineering of energy considerably. For example, when 'socialising' (meeting) is an important and central issue for the project, common rooms will be used more intensively which makes it more suitable for low-temperature heating systems. If shared rooms require servicing, for example with hot tap water, a decision needs to be made to create a separate system or connect to another system already present. The individual servicing simplifies billing but reduces flexibility to adapt the size of units to future household needs.

Differences between projects regarding energy systems are primarily explained by the groups' vision on sustainability, that emphasises one or more of these aspects. For example: Groene Marke, built at initial stages of environmental policies, was motivated to apply the very state of the art and, through self-management, to continue to optimise the energy-system of the project, accepting different spatial-technical standards related to tenure. On the other hand, the Meanderhof group, built during the 'instrumental period' (see Table 10.2) prioritises social interaction and has more distance to technology, emphasising the structural measures reducing the need for energy consumption and avoiding complex technology. IEWAN, built after the stagnation period, bridges the 'DIY' and 'institutional' approaches to achieve a high standard at affordable rents.

A strong involvement of (future) residents during the whole lifespan of the project enables them to influence the energy consumption not only by behaviour, but also by the choice of technology and sources, keeping up with new developments or even own-inventions. Next to sharing equipment or space, building components or utility devices and infrastructure can also be shared in use or in property, for example the common roof may hold a solar installation. The investment may come from the common budget or from more affluent or motivated members who then pay a (modest) fee to the association but receive the larger share of the benefit. Sharing means of transport or combining trips also affects the energy used for mobility.[5] During the design and management, a learning process takes place, which is enhanced through networks. Using the built project as a living lab, outreaching activities take place that are especially interesting for future initiatives, as is illustrated in activities organised by IEWAN. Processes of collective learning and adapted behaviour influence the real performance of the building and technology, and are a precondition to make the hardware function optimally (Baborska-Narozny et al., 2014; Schwenke, 2012).

Finally, in the urban domain, the cluster encompasses the intermediate level between individual buildings and the district. In the energy-household of districts and towns, residential building-clusters have an important role and significant impact on the functioning of energy networks. For example: the peak times of demand for heat in micro- or district heating networks depending on the occupancy of dwellings and flats (commuters-seniors), or the demand and storage of electricity through smart home technology and e-cars. The 'cluster' scale level could be given a specific place in cascading energy flows (such as heat) or rainwater collection, channelling investments between individual utilities and large infrastructures with low-efficiency (high-exergy) values. Until recently, such issues were seen as highly technical, and when a social dimension is involved, it is primarily in terms of 'acceptance' and 'behaviour' (Rahimian et al., 2015; Alber, 2013). This puts the end-user in a passive role, even though changes in behaviour to adapt to new technologies may be expected. At the same, the case studies presented how pro-active citizens with environmental concerns organise as energy-communities, from investing in wind turbines to zero-energy housing. Thus, co-housing initiatives offer opportunities for both the social and technical innovations that the transition towards a renewable energy system requires.

[5] This aspect lies beyond the scope of this paper but strongly influences the preferences of location for co-housing and has significant impact on the overall footprint of projects.

10.8 Conclusions: A Wider Concept of Energy-Communities in Co-Housing

This chapter set out to explore "how far co-housing citizen-led collaborative housing can be considered an energy-community according to the directives' definition?" The research found that co-housing projects qualify as an energy community because of its direct involvement of residents (democratic and autonomous) in decision-making on energy-related practices. Moreover, integrating co-housing is a necessary step to accelerate the process towards a national no-carbon energy-household. Housing is where a large part of generated energy is used (consumed), and increasingly residential buildings are also becoming places where energy is produced. Residents thus become 'prosumers'. This research found that looking at co-housing in this way can open opportunities for housing initiatives to become more sustainable. On the other hand, integrating co-housing is a necessary step to accelerate the process towards a national no-carbon energy-household. However, besides opportunities, the study also identified bottlenecks that pose limitations on decision-making within co-housing regarding energy demand and distribution.

To answer the question 'what contribution could co-housing make to the energy transition?' the research has looked at the engineering of utilities and use of renewable energies in co-housing projects. Residents organised in communities to develop and build their homes often seek low-impact solutions in a holistic way. The initiatives develop energy-related practices that are different from the single standard consumer unit. Combining insights in energy-saving, low-carbon sources and materials, co-housing thus can form an important link in the energy transition. Moreover, the approaches found in co-housing illustrate the interdependency of technical and social innovation, involving design and engineering as well as shared use, involvement in decision-making and pooling resources. Nonetheless, the residents experience sustainability rather as technical, separated from social dynamics. When a technical committee is involved from the very beginning, it accumulates detailed knowledge for maintenance and exploitation. These experiences can be applied in existing clustered housing stock, as well as for the development of isolated off-grid locations.

The qualitative analysis presented here indicates that the possibilities offered by clustered self-managed housing are not fully exploited. While housing collectives are generally prone to going beyond the state-of-the-art in sustainable measures, their success depends on their institutional environment and professional partners. This hardly allow stepwise investment, progressively realising the initial 'energy-neutral' vision instead of discarding it altogether. The lack of general expertise, and legal barriers, force initiatives to rely on classical condominium models. Consequently, the social infrastructure, which could manage, for example, fair billing systems, is also under-utilised. New developments such as apps could be supportive for self-management by providing indicators, benchmarks, comparison between units, signal disruptions or calculations of the billing system.

Besides incidental project figures, there is very little insight into real-time performance. Systematic (comparable) monitoring is rare, and energy performance cannot be calculated with standard models. Consequently, the lessons from co-housing, such as alternative forms of collaboration, management and decision-making can presently not be quantified. The question whether self-managed housing clusters are easily integrated into a decentralised low-carbon energy system is technically answered. The next step needs to be a targeted collection of quantitative data, based on the proposed categories. This contributes evidence to the impact of end-user involvement in 'low-impact' housing and utilities during the lifespan of the project.

Underlying the instrumental lessons, there are also some fundamental questions that need to be raised. For example, can once-off client collectives be expected to have enough knowledge to decide between manifold possible energy concepts when developing the design for their housing cluster? Are consultants equipped to make decisions about technologies that match the ambitions as well as the reality of self-management? Does collective engineering turn housing into exclusive playgrounds for the privileged, with sufficient social capital?

Acknowledgements The researcher is very grateful to the numerous co-housing inhabitants and professionals who shared their knowledge and experiences unconditionally.

References

Ache, P., & Fedrowitz, M. (2012). The Development of Co-Housing Initiatives in Germany. *Built Environment, 38*(3), 395–412.

Alber, G. (2013). Gendered Access to Green Power: Motivations and Barriers for Changing the Energy Provider. In M. Alston & K. Whittenbury (Eds.), *Research, Action and Policy: Addressing the Gendered Impacts of Climate Change.* Springer: https://link.springer.com/book/10.1007/978-94-007-5518-5

Baborska-Narozny, M., Stevenson, F., & Chatterton, P. (2014). A Social Learning Tool – Barriers and Opportunities for Collective Occupant Learning in Low-Carbon Housing. In *Sustainability in Energy and Buildings* (p. 10).

Benson, M. (2014). *Creating a Nation of Selfbuilders. An Interim Report from the Project Selfbuilding: The Production and Consumption of New Homes from the Perspective of Households.* Department of Sociology, Goldsmith. http://www.gold.ac.uk/media/Benson-Creating%20a%20nation%20of%20self-builders.pdf

Buis, H. (Ed.). (2000). *Duurzaam bouwen een kwestie van willen en weten. Voorbeeldprojecten duurzaam en energiezuinig bouwen* [Sustainable and Energy Efficient Building, Report on Demonstration Projects]. Nationaal DuBo Centrum.

Chatterton, P. (2015). *Low Impact Living: A Field Guide for Affordable Ecological Community Building.* Routledge.

Coenen, F., & Hoppe, T. (2018). REScoop Plus: D3.4 – Effectiveness Report 2.

CPB, PBL & SCP. (2014). *Monitor Duurzaam Nederland 2014: Verkenning. Uitdagingen voor adaptief energie-innovatiebeleid* [Monitor Sustainable Netherlands 2014: Challenges for Adaptive Energy Systems].

ECN. (2012). *Energy Efficiency Policies and Measures in The Netherlands. Monitoring of EU and National Energy Efficiency Targets.* ECN, ODYSSEE – MURE. Available from: www.ecn.nl

Equilibre. (2016). Cooperative, Equilibre. "2006–2016 10 ans d'équilibre." Confignons. http://www.cooperative-equilibre.ch.

European Commission Proposes New Rules for Consumer Centred Clean Energy Transition. Available online: https://ec.europa.eu/energy/en/news/

commission-proposes new-rulcs-consumer-centredclean-energy-transition. Accessed 25 May 2020.
European Commission Clean Energy for All Europeans Package Completed: Good for Consumers, Good for Growth and Jobs, and Good for the Planet. Available online: https://ec.europa.eu/info/news/clean-energy-alleuropeans-package-completed-good-consumers-good-growth-and-jobs-and-good-planet-2019-may-22_en. Accessed 25 May 2020.
European Parliament and Council of the European Union DIRECTIVE (EU) 2019/944 OF THE EUROPEAN PARLIAMENT AND OF THE COUNCIL of 5 June 2019 on Common Rules for the Internal Market for Electricity and Amending Directive 2012/27/EU (Recast). Available online: http://data.europa.eu/eli/dir/2019/944/oj. Accessed 25 May 2020.
Fromm, D. (1991). *Collaborative Communities: Cohousing, Central Living, and Other New Forms of Housing with Shared Facilities*. Van Nostrand Reinhold.
Gillinghan, K., Harding, M., & Rapson, D. (2012). Split Incentives in Residential Energy Consumption. *The Energy Journal, 33*(2). https://doi.org/10.5547/01956574.33.2.3
Giorgi, E. (2020). *The Co-Housing phenomenon: Environmental Alliance in Times of Changes* (The Urban Book Series). Springer International Publishing. https://doi.org/10.1007/978-3-030-37097-8
Gram-Hanssen, K. (2013). Efficient Technologies or User Behaviour, Which Is the More Important When Reducing Households' Energy Consumption? *Energy Efficiency, 6*(3), 447–457. https://doi.org/10.1007/s12053-012-9184-4
Gram-Hanssen, K. (2014). New Needs for Better Understanding of Household's Energy Consumption – Behaviour, Lifestyle or Practices? *Architectural Engineering and Design Management, 10*(1–2), 91–107. https://doi.org/10.1080/17452007.2013.837251
Ingle, A., Moezzi, M., Lutzenhiser, L., & Diamond, R. (2014). Better Home Energy Audit Modelling: Incorporating Inhabitant Behaviours. *Building Research & Information, 42*(4), 409–421. https://doi.org/10.1080/09613218.2014.890776
Jarvis, H. (2015). Towards a Deeper Understanding of the Social Architecture of Co-Housing: Evidence from the UK, USA and Australia. *Journal of Urban Research and Practice, 8*(1).
Kido, H., & Nakajima, Y. (2011). *Predicted Energy Conservation by Use of Common Areas in Cohousing*. Kogakuin University. Available at http://www.irbnet.de/daten/icond... Accessed 25 Apr 2013.

Krokfors, K. (2012). Co-Housing in the Making. *Built Environment, 38*(3), 309–314.

Lang, R., & Stoeger, H. (2017). The Role of the Local Institutional Context in Understanding Collaborative Housing Models: Empirical Evidence from Austria. *International Journal of Housing Policy*. https://doi.org/10.1080/19491247.2016.1265265

Locatelli, D., Desrues, F., & Biry, J.-M. (Eds.). (2011). *Guide Pratique de l'Auto-promotion*. Association Eco-Quartier Strasbourg et CAUE Bas-Rhin.

Marckmann, B., Gram-Hanssen, K., & Haunstrup Christensen, T. (2012). Sustainable Living and Co-Housing: Evidence from a Case Study of Eco-Villages. *Built Environment, 38*(3), 413–429.

Palojärvi A., Pyysiäinen J., & Saloranta M. (Eds.). (2013). *Inspiring Stories from Ecovillages: Experiences with Ecological Technologies and Practices*. www.balticecovillages.eu

Rahimian, M., Domenica Lulo, L., & Cardoso Llach, D. (2015). The Case for a Collaborative Energy Sharing Network for Small Scale Community Microgrids. In *Architecture and Resilience on the Human Scale* (pp. 193–200). University of Sheffield.

Rescoop. (2016). URL: https://rescoop.eu/european-project/community-power. Accessed 2016.

Remkes, J., & Pronk, J. (2000). *Mensen Wensen Wonen. Nota wonen in de 21ste eeuw*. Ministerie VROM.

Roetgerink, F. (2006). *Bouwen aan effectief beleid voor particulier opdrachtgeverschap*. Policy recomendations. i.o.v. Min. VROM DGW. TU Twente, Faculteit BTT, July 12, 2006.

Sargisson, L. (2012). Second-Wave Cohousing: A Modern Utopia? *Utopian Studides, Pennsylvania State University Press, 23*(1), 28–56.

Schwenke, A. M. (2012). *Energieke BottomUp in Lage Landen. De Energietransitie van Onderaf*. AS I-Search. Available from: http://www.asisearch.nl/energieke-bottomup-in-lage-landen. Accessed 21 Sept 2012.

Stevenson F., Baker H., & Fewson K. (2013). *Cohousing Case Studies in the UK: Is Sharing Facilities Really Resourceful?* PLEA2013 Conference Proceedings. Munchen.

Stofberg, F., Gommans, L., & Israëls, E. (2000). *Vademecum energiebewust ontwerpen van nieuwbouwwoningen. 2de druk*. NOVEM.

Subrémon, H. (2011). *Anthropologie des usages de l'énergie dans l'habitat*. PUCA.

Sundberg, F. (2014). *Cohousing and Resource Use a Case Study of the Färdknäppen Cohouse*. Master thesis, Kungliga Tekniska Högskolan School of Architecture and the Built Environment.

Tillie, N., van den Dobbelsteen, A., Doepel, D., de Jager, W., Joubert, M., & Mayenburg, D. (2009). Towards CO_2 Neutral Urban Planning – Introducing the Rotterdam Energy Approach and Planning (REAP). *Journal of Green Building, 4*(3), 103–112.

Tummers, L. (2013). Co-Housing: Pioneers of Eco-Engineering. In *Overcoming the Crisis, Integrating Urban Environments*.

Tummers, L. (Ed.). (2015a). *The Re-Emergence of Co-Housing in Europe*. Routledge. https://www.routledge.com/products/9781138961258

Tummers, L. (2015b). The Re-Emergence of self-Managed Co-Housing in Europe a Critical Review of Co-Housing Research. *Urban Studies, 23*(May), 1–18. https://doi.org/10.1177/0042098015586696

Tummers, L. (2017). *Between Passive House and Active Residents: Learning from Co-Housing Initiatives*. ABE 14. Delft: Delft University of Technology. https://journals.open.tudelft.nl/abe/article/view/1858.

Tummers, L. (Guest ed.), & Palmer, J. (Guest ed.). (2019). Collaborative Housing: Resident and Professional Roles. *Built Environment, 45*(3). https://www.ingentaconnect.com/content/alex/benv/2019/00000045/00000003#expand/collapse

van den Ham, M., & Keers, G. (2010). *CPO: markt voor bouwers*. Rigoin opdrachtvan St. RRBOUW, November 2010. rigo.nl.

van der Reijden, H., Leidelmeijer, K., & Marsman, G. (2002). *Gebruikerservaringen Voorbeeldprojecten Duurzaam & Energiezuinig bouwen Woningbouw*. Rigo voor SEV & Novem.

van der Woude, H. (2012). *Community Architecture in Nederland*. Toth.

Van Voorden, de Groot Groep, & De Regie. (2004). *Onderzoek kosten- en kwaliteitsvergelijking collectief partikulier opdrachtgeverschap. SEV programma De nieuwe bouwopgave*. SEV.

Vestbro, D. (2000). From Collective Housing to Cohousing – A Summary of Research. *Journal of Architectural and Planning Research, 17*(2), 164–177.

Vestbro, D. (Ed.). (2010). *Living Together- Co-Housing Ideas and Realities Around the World*. Royal Institute of Technology Division of Urban Studies in Collaboration with Kollektivhus NU.

Williams, J. (2005). Designing Neighbourhoods for Social Interaction: The Case of Cohousing. *Journal of Urban Design, 10*(1), 195–227.

11

Conclusion

Frans H. J. M. Coenen and Thomas Hoppe

11.1 Introduction

In this volume, the key questions were: how renewable energy communities in continental Europe have evolved, and what can be learned from these experiences for the realisation and implementation of the Clean Energy Package? In the previous chapters of this volume authors have addressed these questions in one way or another.

The Clean Energy for All Europeans Legislative Package (CEP) was concluded by the European Union institutions in May 2019. The CEP

F. H. J. M. Coenen (✉)
Faculty of Behavioural, Management and Social Sciences, University of Twente, Enschede, The Netherlands
e-mail: f.h.j.m.coenen@utwente.nl

T. Hoppe
Faculty of Technology, Policy and Management, Delft University of Technology, Delft, The Netherlands
e-mail: t.hoppe@tudelft.nl

© The Author(s), under exclusive license to Springer Nature Switzerland AG 2021

F. H. J. M. Coenen, T. Hoppe (eds.), *Renewable Energy Communities and the Low Carbon Energy Transition in Europe*, https://doi.org/10.1007/978-3-030-84440-0_11

concerns a legal framework that will help the EU to meet its 2030 climate mitigation and clean energy policy goals. As Roberts explained in Chap. 2 with this legislative package, the EU signalled a strong shift in the role of citizens from passive consumers to active participants in energy transitions. This marked the first time that EU legislation acknowledged the role community energy ownership can have to help the EU meet its climate and energy goals while driving local social innovation at the same time. In particular, the recast Directive 2018/2001 (Renewable Energy Directive II, or REDII), the recast Directive 2019/944 (the Internal Electricity Market Directive, or IEMD) and the recast Regulation 2019/943 (the Internal Electricity Market Regulation, or IEMR) contain provisions that establish a supportive EU legal framework for community ownership. Other EU actions follow the same idea of giving citizens a more central position in the energy transition. This particularly holds for the EU Strategy for Energy System Integration (2020) and for elements from the European Green Deal package.

The four sub-questions central to this volume relate to a number of observations that are discussed in the Introduction chapter of this volume. All chapters following the Introduction chapter either focus on the developments around renewable energy communities in a certain country, like Greece, Poland, Estonia, Latvia, Lithuania, Estonia, Italy, Germany and the Netherlands, or alternatively address a certain aspect of the energy community's strategy, involvement or actions and relevant underlying ideas. There is also attention to countries that until recently received hardly any or no attention in the community energy literature. These include Italy, Greece, Poland and the Baltic states of Estonia, Latvia and Lithuania. The set of countries presented in this volume is however by no means representative for regional clusters of countries, but they are either part of the front-runner group of countries, namely the Netherlands and Germany, or part of the late-starters group, namely Southern and Eastern European countries.

The CEP defines two new concepts labelled 'renewable energy communities' (RECs) and 'citizen energy communities' (CECs). It also requires Member States to secure certain rights of energy communities and establish enabling frameworks to ensure a level playing field and promote their development. EU Member States were requested to implement REDII provisions into national legislation by 30 June 2021, and the IEMD provisions by 31 December 2020, to ensure that they are consistent with the new EU legislation.

In this conclusion chapter, we refer to Renewable Energy Communities, as RECs. For the sake of clarity, REC and CEC are not mentioned separately. What holds for RECs in practice often also holds for CECs.

11.2 Instrumental Functions of Community Energy

In general, non-profit and voluntary organizations can serve an instrumental purpose in meeting society's needs. This is what many founders of community energy initiatives had in mind when they started their initiatives. Over the last thirty years, their idealistic aim was in the first place to contribute to combating climate change by producing renewable energy. However, in many cases this was not the only goal as they also pursued other value-oriented goals, not all of them addressing the ecological dimension of sustainable development. Other goals pertained to showing anti-government, anti-centralist, multinationals-led energy markets, or anti-nuclear energy motives (Hess, 2018). This can be seen as a critical civic response to centralist, capitalist, and eventually developments that are considered harmful to the environment and have restricted the autonomy of local and regional communities (Smith et al., 2016; Hewitt et al., 2019).

In some sense, this can also be seen as a response to the ways in which the energy liberalisation across EU Member States failed to effectively address the greening of energy systems by giving market players more influence. It also neglected the interests of citizens and local communities and payed attention to other values rather than limiting energy market monopolies and assuring low, affordable energy pricing. More in general, civic response to perceived market and state induced abuses is typical for the way social innovations develop. This particularly holds for community energy (Hewitt et al., 2019).

Moreover, the community energy movement can also be seen as having a political side, seeking to regain more decentralised power into governing and organizing energy supply chains, and taking decision-making power away from centralist incumbent players, in particular national governments and the so-called 'energy giants' (i.e., the incumbent energy companies that form oligopolies in domestic energy markets). Some energy communities even go so far that they desire to establish local autarkic self-sufficient energy systems that operate fully independent from national and regional energy system (e.g., 'Klimakommüne Saerbeck'; Hoppe et al., 2015). More in general, it can be observed that the community energy movement seeks to achieve a more polycentric model of energy supply systems (Bauwens, 2017) as an alternative to the centralist, monocentric model in which energy markets have been organised over the last century, although the liberalisation to a certain extent can be seen as another—yet less far-reaching—way to establish more polycentric energy supply systems.

Previously, dating as far back as the end of the nineteenth century as reported by Candelise and Ruggieri in Chap. 5 of this volume, energy communities in the form of cooperatives started to provide rural areas with electricity due to a lack of national grids and a lack of prioritisation of these rural areas. The same happened in Germany as described by Holstenkamp in Chap. 6. Internationally this was also observed in countries like the US (Yadoo & Cruickshank, 2010). The instrumental purpose in meeting society's needs here contributed to the rural economy and increasing rural welfare. Access to electricity is crucial in rural development all around the world (World Bank, 2017). Access to affordable and clean energy is not only a key element of the Sustainable Development Goals (SDG7) but also of the social equality of women in energy transition as discussed by Feenstra and Hanke in Chap. 9.

The CEP stresses how important the contribution of individual citizens and citizen's community initiatives are in relation to the issue of sustainable energy transitions. Feenstra and Hanke (see Chap. 10) argue that in addition this should also be a socially inclusive energy transition. More recently, a leading role for citizens is observed in drawing attention for climate change and addressing power structures through (social) media and direct action by civic action groups like Extinction Rebellion

(Westwell & Bunting, 2020). On the other hand, citizens are also observed to oppose energy transitions, its costs and consequences for the economy and individuals (with increased tax rates and the introduction of transport levies which for instance sparked the yellow jackets' civic resistance movement in France; Jetten et al., 2020). Moreover, in particular this addresses public acceptance of the siting of renewable energy projects (Devine-Wright et al., 2017; Wüstenhagen et al., 2007; Lennon et al., 2019). Some fear that renewable energy, although sustainable, is no longer secure nor affordable. Feenstra and Hanke (see Chap. 10) place this in context of the awareness of RECs of energy poverty and gender inequality. Others fear that the present energy market and competition will not lead to sufficient amounts of renewable energy becoming available (Paravantis & Kontoulis, 2020). Some even vie for an increased share of nuclear power into the energy mix, framing it as a clean low carbon technology.

Mobilising and getting citizens involved in sustainable energy transitions for the sake of progressing the latter can be perceived as taking a fairly reductionist instrumental perspective (Wittmayer et al., 2020). From a normative perspective community involvement and more particularly community energy action is considered as positive to citizens themselves, taking a citizen empowerment or from a democratic values perspective (i.e., pertaining to energy democracy goals) (Van Veelen & van der Horst, 2018). It is instrumental because it contributes to facilitating the processes and action towards achieving goals that are related to sustainable energy transitions. Although democratic values are still important the Clean Energy Package also highlights the aspect of distributional benefits that come along with energy transitions. These so-called 'co-benefits' for citizens and (local) communities are instrumental because they might spark citizens into making investments, which in turn may unlock private capital for renewable energy production investments. However, one also has to realise that citizen investments can be made from a personal gain perspective and not only deriving from idealistic, ecological-collectivist motives. Individual citizens might even view solar energy as a better investment than investing in most shares on the stock exchange. Their contribution to climate change is for them just a side effect or sub-goal. In the Ecopower case Bauwens (2016) showed that new cohorts of people joining REScoops are less idealistically motivated than the ones that started these initiatives.

In Chap. 1 of this volume four observations of general developments in community energy were discussed. The first observation made pertained to the increase of importance of the instrumental function of citizens' involvement in energy transitions. Based on this observation, the following sub-question was formulated, *What does the general trend of a greater focus on a more instrumental function of community energy for the EU energy transition mean for the other more normative functions of renewable energy cooperatives?* This led to the underlying issue addressing whether citizens' involvement in sustainable energy transitions actually contributes to reaching energy transition's goals. In Chap. 10 Tummers shows the possibility of citizens' communities contributing to a stronger energy performance in housing than traditional housing projects. In Chap. 8 Hoppe and Coenen show the potential contribution of energy communities to energy savings by their members based on community membership and actions, social norm-setting, knowledge provision, as well as personal beliefs and attitude. In Chap. 2 Roberts addresses the requirement that RECs focus on providing economic, environmental or social benefits. Although issues like these seem to be addressed in the CEP with the aim to have them implemented into national legislation it appears that they are not clearly operationalised. In general, most attention on benefits concerns environmental benefits of RECs through increased production of locally produced renewable energy and greenhouse gas emission reductions.

The question can be raised whether *the tendency towards an instrumental perspective of community energy can influence the value and principles of its cooperative identity?* The requirement that RECs focus on providing economic, environmental or social benefits also includes typical benefits that can be perceived as previously mentioned, as "a more normative function of RECs". For example, an important part of the social benefits to RECs concerns their promotion of energy democracy, citizen empowerment and rising the general awareness on climate issues of citizens and REC members. This last point was explicitly addressed in Chap. 8 by Coenen and Hoppe on RECs encouraging their members to engage in energy-saving actions. In Chap. 2, Roberts shows that the revised

EU directive contains many requirements for EU Members States to implement into national (and sub-national) legislation to empower and protect the ways in which RECs are functioning; in the first place because the legal identity in which RECs must be organised around specific ownership and governance principles, having non-commercial purposes. In addition, specific requirements are in place to check that these basic principles are upheld, and prevent RECs from becoming public utilities or private energy companies under disguise.

However, there is a potential tension between the original idealistic motives of saving energy and producing renewables, and becoming serious players in energy markets. A call for scaling and the professionalisation of RECs due to market requirements leads to staff professionalisation and development of new, more professional business models, and potentially a larger distance to the community, despite the fact that there are certain specific requirements in RED II in place to facilitate the market position for RECs without losing the REC identity in terms of citizens ownership, democratic control, local autonomy and having non-commercial purposes.

A growing involvement is observed of other actors than citizens in community energy. Of course, local governments and other intermediaries have traditionally been involved and have supported community initiatives (see also Chap. 7 by Warbroek). In addition, many municipalities, profit and non-profit organisations (Caramizaru & Uihlein, 2020; Warbroek & Hoppe, 2017; Hufen & Koppenjan, 2015) have actually started community energy initiatives themselves (see the next paragraph). The definition of shareholders or members that are located in the proximity of the renewable energy projects owned and developed by renewable energy communities as legal entity opens membership of RECs to others than just citizens and citizen groups. Therefore, some authors might want to reduce community energy to just citizen's initiatives as for example presented in Chap. 3 by Ruggiero et al. This might spur a discussion on what a community exactly entails and what role geographical proximity plays in the idea of community. The main argument from the EU for geographical proximity criteria for members and stakeholders is the substantial added value in generating local acceptance of renewable energy projects. Some REScoops are (national) umbrella organisations with local groups or members. Or they are local groups scaling and spreading to other places as described by Candelise and Ruggieri in Chap. 5. In response to the implementation of RED II it may be expected that these broadly oriented and large-sized REScoops will be subdivided into smaller organisations or local groups that meet the proximity criteria of art. 2 of RED II.

When reflecting on the trend towards this functionality perspective on RECs it is important to address who started RECs as a social initiative in the first place. In Chap. 3 Ruggiero cs. mention that in Poland the driving force behind the establishment of energy clusters are local authorities and business companies with a low level of citizen engagement. In Chap. 5 Candelise and Ruggieri discuss to which extent REC initiatives in Italy have been proposed and developed by citizens or other types of grassroots organisations (bottom-up), or instead by a municipality that defines the project and the forms of citizens' involvement; it turns out that the majority of the Italian initiatives have been initiated using a top-down structure, with five of the REC initiatives having been proposed by local government, and the other seven by a (semi-)commercial actor (i.e., either a company or a municipal utility). Only five REC initiatives were actually initiated via a bottom-up approach by either a group of citizens or a green NGOs. In Chap. 4 (by Ziozas and Tsoutsos), in Greece the two REC cases from the 1980s and 1990s were initiated respectively by a power company and a municipality. The three newer Greek REC cases pertain to two citizen-led energy initiatives and one by an intermediary actor dedicated to the development of local energy projects.

Does a strong involvement of municipalities and particular private actors mean that RECs become less idealistic local community initiatives? Here the autonomous and effectively controlled criteria as explained by Roberts in Chap. 2 are important. The REC definition states that it must be "*effectively controlled by shareholders or members that are located in the proximity of the renewable energy projects that are owned and developed by that legal entity*." As explained by Roberts, proximity should be generally understood as the geographical scope in which the members or shareholders that effectively control the REC should be located (in other words: reside). Here, geographical proximity is emphasized. Some countries like Belgium or the Netherlands have national organizations or federations that function as a kind of umbrella organisation or branch organisation for RECs. The question in these cases is at which organisational level crucial decisions are taken and criteria set. Membership democracy in such an organisation would imply indirect democracy.

The REC definition explicitly mentions the proximity criteria and benefits criteria. In Chap. 2 Roberts explains the primary purpose of REC to provide environmental, economic or social community benefits to its members or shareholders or to the local areas where it operates rather than to generating financial profits. This restricts real renewable community initiative to not-for-profit and community initiatives.

Another way that RECs keep their more idealistic and normative principles is by ensuring that a specific category of members (e.g., natural persons or local authorities) are included in energy communities. Keeping citizens as natural persons in the lead in these initiatives follows from what we called the normative idea in the energy community concepts aiming at empowering citizens to participate and take ownership. In Chap. 9 Feenstra and Hanke advocate a special position for women to be able to engage in RECs on an equal basis.

Implementing such a special position is potentially feasible because in national regulation following RED II it should be ensured that citizens have a minimum level of participation or a decisive influence in the communities that are formed. This minimum level of participation or decisive influence will only lead to the uptake of the more idealistic goals if the members share these goals. The conditions are that individual citizen members, for example, should not view joining an REC as an opportunity to obtain solar energy as just a better investment than investing in stocks. These members are considered to also be interested in actively participating in the activities and strategic decision-making of RECs.

On the other hand, if RECs do not also function as an ideal democracy of motivated energy citizens they can still play an important role in energy transitions. In Chap. 3 Ruggiero cs. hold that despite the limitations of the Polish energy cluster model, these clusters can serve as an opportunity for CE groups and become a vehicle for the phase-out of coal in the long-term future. And as Tummers shows in Chap. 10 that although self-managed housing initiatives do not use the original model RECs (or as cooperatives) are be considered as more than a group of jointly consuming prosumers, for they can also play an important role in showing what is possible in the housing sector when it takes more of a REC, collectivist, ecological, democratic value-oriented approach.

11.3 Different Starting Positions of EU Member States

The second observation discussed in Chap. 1 is that there are clear differences between front-running and late-starting countries in their original position. Nevertheless, they currently face the challenge of implementing with the same directive. As Roberts discussed in Chap. 2, transposing these new EU rules into national legislation is considered as challenging.

The second sub-question based on this observation is: *What does the new EU policy mean for countries that are late-starters or front-runners in the community energy movement?*

Continental Europe by no means presents a homogeneous landscape when one looks at the deployment and development of renewable community communities. In Chap. 1 it was discussed that there are a number of countries that would fall in the category of front-runners. Within this group of countries, this volume gives particular attention to Germany and the Netherlands. Of the relative late-starters within the groups of Southern European countries particular attention was given to Italy and Greece. Within the Eastern European countries particular attention was awarded to Poland and the Baltic States Estonia, Latvia and Lithuania. Eastern European countries are not homogeneous but share a number of characteristics. For the Baltic States, this is Soviet heritage. For Poland and other Eastern European countries, the previous communistic regimes created a different starting position. First, as Ruggiero cs. point out in Chap. 3 the repression of an independent civil society by the Soviet regime and previous communistic regimes has led to a situation in which civil and especially collective activism (i.e., mostly in the form of energy cooperatives) is still less developed in Eastern European states than in their Western counterparts (Aidukaiu, 2013). Second, the often-times negative experience of forced collectivisation of farms has created a level of distrust towards community projects. This might even lead to an aversion of for using words as energy cooperative or community energy. The two other countries in this volume, Italy and Greece, which represent another group of relatively late-starting countries, have a different starting point in this perspective. Cooperatives are a well-known and valued organisational forms in both countries, in particular in agriculture. They also have a tradition in having an independent civil society, which is however relatively weak in REC development in Greece according to Ziozas and Tsoutsos (See Chap. 4). If we compare the two Southern European countries with the Baltic and Eastern European countries, in Italy the first energy communities date back to the end of the nineteenth century and in Greece, there are 'historical' cases for the 1980s and 1990s. Bauwens, Gotchev and Holstenkamp (2016) identify several important factors that are intended to clarify the disparities between EU countries, such as formal institutional rules, the support mechanisms for renewable energy production and spatial planning, attitudes toward the cooperative model and the cultures of local energy activism. These factors seem also to have explanatory power for the countries that are less experienced with RECs.

For the Eastern European countries in particular the general lack of civic engagement as essential for starting RECs (Walker & Devine-Wright, 2008; Radtke, 2014) explains their late-start. In comparison in the Southern European countries; this is more specific local renewable energy activism. The other factors, influenced by the communist heritage, are low trust and fairly low levels of social capital as important driving factors for citizens (Walker & McCarthy, 2010). Moreover, characteristic to the Eastern European countries is having a negative attitude towards the cooperative business and legal-organizational model.

For all countries in the group of 'new' countries, economy plays a role. The Eastern European countries and the Greek economy are weaker than the economy of the front-running countries. However, more important than the general economy are the consequences for the financial position of individual citizens and the financial position of municipalities. Higher levels of income and education, which are often inter-related, are also a positive influence factor for starting a REC (Radtke, 2014; Magnani & Osti, 2016).

Another contextual factor also mentioned by Ruggiero cs. in Chap. 3 is decentralisation of authority and resources to well-functioning sub-national governments. If there is decentralisation but the institutional capacity is limited the influence of decentralisation is reduced. However, the Netherlands is considered as one of the front-running countries with well-functioning local governments that however lack authority, have a low independent tax-income and tend to increasingly start lacking administrative capacity. Chapter 7 by Warboek shows that intermediaries can support the 'weaker' municipalities (i.e., small-sized, with low capacity, and with underprivileged socio-economic communitites).

In Chap. 4, Ziozas and Tsoutsos explain what they call "*the enormous gap between local communities and RES projects*" as a result of the limited governmental and economic freedom that municipalities and communities experience in Greece due to the centralised State structure. On the other hand, in the United States there is still a considerable number of cases reported (Van der Schoor & Scholtens, 2021) although there is a relative absence of federal energy policies.

The national economy strongly relates to the energy market and particular energy prices. Low fossil fuel-generated energy prices on the national electricity markets can act as a barrier to making REC projects economically viable. This also relates to dominance of incumbent energy companies in the national economy (Proka et al., 2018; Lowes et al., 2020) and the energy sector and the favouring of a certain form of fossil energy for economic and political reasons like the importance of coal in Poland and biomass in Latvia.

Finally, regulation has to be in place. In Chap. 3 Ruggiero cs. address the importance of feed-in tariffs, but also costs of grid connection, high interest rates for loans and the government support for receiving loans. In Chap. 3 Warbroek points to the role and strategies of intermediaries in building capacities of RECs, embedding them into the community, alleviating barriers and opening up the system for the uptake, acceptance or breakthrough of RECs. In Chap. 5 Candelise and Ruggieri show how generous feed-in tariffs positively influenced the first wave of RECs in Italy but the disappearance also caused a hold on the growth of RECs. The importance of having feed-in tariffs in place to empower RECs was also observed in other predominantly North-Western European countries (Wierling et al., 2018).

The transportation of the CEP into national regulation is expected to change many regular and financial conditions in the European countries. However, this volume was published too early to give a clear, and complete answer to the question how countries faired in adopting regulatory frameworks at the national level.

The CEP and the regulations, and regulatory change that follow from it make a clear promise for a number of issues that are essential to the survival and upscaling of RECs. However, some of these issues depend on internal organisational problems RECs have and will not be solved by national regulation (see Chap. 7 by Warbroek) like problems with capacity and knowledge of volunteers, tensions with professionalisation and financing. RECs are not like the other market players due to their principles on participation, ownership and decision-making. This has a number of consequences in the way they operate in the energy market and create and run energy projects. Because their profits do not go to shareholders, but are reinvested in benefits to the community their return on investment is often lower than commercial parties running renewable energy projects. Raising funds and managing to develop or adopt feasible business cases (i.e., solving unprofitable top issues) is also more complicated for energy communities, especially when it comes to risky prefinancing that is necessary, while the continuation of the project is not certain. Moreover, having a fairly 'alternative' (i.e. non-conventional) organizational form does not make it any easier to get funding and obtain sufficient loans from banks. Furthermore, due to their democratic principles decision-making processes are more complex and therefore slower to act than for commercial actors. In addition, the existing problems with working with dedicated volunteers who do not always have the necessary legal expertise in the context of permits or licences is increasingly becoming a problem in opening up energy markets for RECs.

11.4 Broadening of Goals and Activities of Renewable Energy Communities

The third observation made in Chap. 1 concerned the trend of broadening the goals and activities of RECs over the years. The third sub-question is: *What does the broadening of their goals and activities mean for renewable energy communities and what role does the directive play?*

Many examples in this volume are about community-owned wind turbines, solar parks or other solar projects on for instance public buildings and biomass community projects. There are RECs involved in district heating based on biomass or as multi-utility cooperatives jointly producing renewable energy, heat next to power (Coenen & Hoppe, 2018). There is a history of energy cooperatives also producing district heating (Vansintjan, 2015), or as truly embedded systems producing heat and electricity through renewable biomass and combined heat and power (CHP) systems, with a local heat network delivering heat to households (Yildiz et al., 2015). Exceptional are RECs involved in the financing of hydro power plants or investing in biogas production (Som Energia, 2021). The EU is applying the principle of decentralisation and citizen-centred transformation from the CEP to a larger area of policy proposals that are being put forward under the European Green Deal. What will become particularly important for RECs in broadening their goals is that they enter other ways of energy production than just producing renewable electricity (i.e.,, the EU Strategy for Energy System Integration, 2020).

Paragraph 3 of Article 2 of the revised Renewable Energy Directive means an explicit recognition that community energy is not just about jointly producing renewable energy, and provides legitimacy for RECs to broaden their activities. Besides energy generation, including from renewable sources, distribution, supply, consumption, aggregation, energy storage, and a number of other activities are explicitly mentioned (in particularly to CECs). This includes engaging in energy efficiency services or charging services for electric vehicles; or to provide additional energy services to its members or shareholders. In Chap. 8 Hoppe and Coenen discuss the role energy communities can play in energy saving and why they are very well placed to do so. In Chap. 5 Candelise and

Ruggieri indicate that there are already RECs in Italy that focus on broader tasks than just producing renewable energy.

Many examples of energy-saving activities of the RECs towards their members and the broader community pertain to examples of a new type of REC that are becoming service providers of more energy services than just supplying renewable energy to end consumers. This however also requires change within RECs that have to enter new networks and see the value of an integrated system. In addtion, it also requires RECs to possess sufficient knowledge and capacities on legislation within other areas like building codes. And this legislation is also changing. New EU legislation will be transposed into new national regulations and eventually new local rules. An example is an REC getting into the electrification of transport by organising e-car sharing (Schwabe, 2020; Bocken et al., 2020). This is an example of the production of renewable energy for the organisation of charging points, sustainable transport but also the buildings, where the charging points are often located.

11.5 New Insights Observed in Late-Starting Countries

The fourth observation introduced in Chap. 1 concerned a mismatch between academic attention and significance of the community energy movement in countries. A substantial part of the community energy literature focuses on the UK. The current academic literature predominantly uses cases from the UK with contributions from a few other countries such as Germany, Denmark and the Netherlands. In addition, whereas there are numerous studies analysing community initiatives in Central and Western Europe, there is far less research on community energy in Eastern European countries (Ruggiero et al., 2021). The same holds for Southern European countries.

For the questions in this volume, it is not relevant from which country the authors come but of which countries the empirical material is obtained. Based on an extensive review of literature on community energy Van der Schoor and Scholten (2021) conclude that there is a substantial geographical bias in community energy studies. They looked at the geographical distribution of cases because the majority of the papers in the community energy literature rely on a case study approach and focus on one or more specific countries. Given that some studies compare cases in two or more countries, no less than 308 cases were analysed. Of these cases, one-third are about community energy in the UK, with Germany, the Netherlands and the US ranking second, third and fourth. Combined, more than two-thirds of all case studies on RECs study community energy in one of these four countries.

It is not easy to explain why less research on RECs is undertaken in other countries. One would first expect that there needs to be substantial cases before they can be researched and described, which might be a problem in the countries that are lagging behind. Another factor would be the size and intensity of social scientists studying community energy, which might explain for the attention in some smaller countries like Denmark and Austria compared to the UK. However, this does not explain for the position of the Netherlands as a smaller country. Another logical explanation would be the significance of the climate discussion in a country and the significant share of renewable energy from community energy projects. In Germany, RECs are responsible for a considerable part of the renewable energy produced. In contrast, if one looks at the significance of local renewable energy and strong renewable energy polices the lack of research on Demark is remarkable (Van der Schoor & Scholtens, 2021). This brings us to the fourth sub-question: *Does more explicit attention to continental Europe and particular late-starting countries bring new insights?*

There a difference between countries that did get less attention in the academic literature, countries with less disseminated renewable energy communities and countries where the development of RECs started later. Chapter 5 by Candelise and Ruggieri on developments in Italy and Chap. 4 by Ziozas and Tsoutsos on developments in Greece show that there were early cases of REC activity in Italy and Greece. However, these appear lacking in Eastern European countries. Although the development and impacts of community energy undertakings are quite dependent on local cultural and political conditions and policies, from the chapter contributions in this volume it looks that many factors are the same for the front-runners and late-starting countries.

Barriers for starting RECs were found to be roughly the same in the front-running and later-starting countries, but some barriers seem to be specific or have a stronger presence in the latter. In Eastern European countries, the dislike and distrust of cooperatives and communities and the lower level of civic engagement are a specific barrier. The problem of a lower civil society engagement is also present in Italy and Greece, but more specifically for renewable energy projects. Looking at other countries that receive less attention in the academic literature, they point us to the large variety of legal and organizational forms of collective renewable energy prosumers (Horstink et al., 2021). The case of Poland shows stronger involvement of private sector partners and municipalities, and initiatives from their side due to lower civic engagement.

11.6 Discussion

This volume began by raising the following question: *How did renewable energy communities in continental Europe evolve, and what can be learned from these experiences for the realisation of the Clean Energy Package?*

The CEP eventually will create conditions for better functioning energy markets with regard to the position of RECs, and in particular with an improved level playing field. However, there are also many factors in energy markets where RECs operate that still might influence their functioning. Moreover, even if the supportive EU legal framework for RECs is implemented into national legislation and policy frameworks, not all conditions for RECs will be optimal. There are still many factors grounded in the general economic situation, the broader energy markets, the attitude of citizens and the effect of the history and culture in countries that will be difficult to change. And there are developments outside of the EU that will influence CEP implementation, like national factors related to technological, economic and domestic energy market developments. Most of the experiences we discussed in this volume were with renewable energy communities and their predecessors. What problems does the CEP actually change for the existing and new RECs, and where does one run into structural problems that cannot be solved easily? In this section, the realisation of the CEP will be discussed from the perspective of the successful translation of the principles of the CEP into national regulation and policy frameworks, along with potential problems that need to be overcome that hinder this process. This includes taking the perspectives of energy markets, business models and technological development.

First, the question can be raised as to how far the realisation of the CEP depends on the successful translation of the principles of the CEP into national regulation and policy. The new electricity market rules that the European Union put forward and need to be translated into new national laws that give consumers the right to produce, sell and share their own electricity in renewable energy communities. The chapters in this volume cannot provide clear answers to the question how the implementation of the directive itself evolves. According to Holstenkamp (see Chap. 6) Germany did not make the deadline for transposing the recast Renewable Energy Directive (RED II) into national legislation. The German legislator had at the moment of writing (2021) not taken any step to build a definition of energy communities and further support structures into the German Renewable Energy Sources Act. In contrast, late-starter Greece was fairly early with developing regulation that accommodates CEP with the 2018 law on energy communities,[1] which adopts cooperatives as the basis for its definition of energy communities but still a difficult situation for starting RECs (See Chap. 4 by Ziozas and Tsoutsos). At the moment of writing, in the Netherlands the bill concerning the new Energy Act that will implement the CEP was only recently finalized after public consultation. The Dutch energy cooperatives generally meet the European definition(s) (see Chap. 7 by Warbroek). However, the Dutch community energy movement was fairly disappointed with the delayed transposition of the CEP into national legislation. Moreover, RECs will likely not be adopted nor occur in the "new heating Act".

A number of EU Member States already have some regulations and policy in place to support smaller and non-commercial market actors in the energy market as well as more decentralised renewable energy production. For EU Member States that already have elements in place the transposition of the EU Directives offer an opportunity to upgrade and

[1] Law 4430/2016 on Social and Solidarity Economy and the development of its actors.

expand, but as mentioned by Holstenkamp in Chap. 6 this does not necessarily happen in the front-runner countries yet. The author argues that he does not observe a "revival" of community energy after the introduction of the CEP. As far as the CEP is a "window of opportunity" for RECs and citizen energy, this opportunity was not taken in Germany. In contrast, in the Netherlands cooperative communities were observed anticipating on the introduction of the CEP by introducing the concept of a thermal energy community (*warmteschap* in Dutch; Schwencke, 2021; see Chap. 7 by Warbroek).

The importance of countries having feed-in tariffs (FiT's) to support the evolvement of RECs was addressed. However, the EU decided to abandon this economic incentive and adopt a more market-driven approach to boosting renewable energy. In 2014, the FIT's, which had supported the citizen projects, were considered as 'permissible State aid'. However, the 2014 EU Guidelines on State aid stipulated that auctions should be the preferred policy for renewable electricity. The assessment of the progress of electricity generation from wind and solar photovoltaic power by the European Court of Auditors (2019) led to the recommendation to the Commission to help EU Member States support further deployment by organizing auctions. However, auctions and tenders raise high thresholds for citizen initiatives. It makes it harder for RECs to enter energy markets on a fair level playing field. That is for instance because they cannot spread financial risk across different projects. In Chap. 6 Holstenkamp discusses problems with the German wind auction project where commercial project developers misused rules. Since 2016, Greece has moved from a standard FiT system to a tender process to support renewables. In Chap. 4 Ziozas and Tsoutsos mention a very recent Greek Law 4759/2020, which obliges all RECs from 2022 to compete with private investors in bids to ensure the operational support of RES projects.

How important is the EU by setting framework conditions to secure certain rights of RECs and establish enabling frameworks to ensure a level playing field and promote their development? For the EU internal market it is important that rules in Member States are the same. The stepping stone for the CEP was that in the past Member States had adopted different national frameworks. This was assessed in the process leading up to the CEP and in revising the directives. The previous directive RED I had already offered the possibility to regulate renewables' self-consumption. However, there was a fragmented situation in national legislations that were nevertheless compliant with RED I, i.e., allowing prosumers to produce electricity for self-consumption, although in some countries there were barriers in the form of necessary licences and incurred costs. For instance in Poland, prosumers faced barriers like taxation and limitation on the benefits from electricity fed into the electricity grid (European Court of Auditors, 2019).

Two main changes that RED II introduces are the concept of renewables self-consumers and jointly acting self-consumers. The aims of RED II are to enable and regulate their contribution to the development of renewable energy. The EU Member States are obliged to include in their national regulation that self-consumers are not to be subjected to excessive, discriminatory charges or fees in relation to the electricity consumed, generated, or fed into the grid. However, in some cases, EU Member States may apply non-discriminatory and proportionate charges and fees to renewables self-consumers. They can differentiate between individual and jointly-acting renewables self-consumers in a proportionate and duly justified way.

A number of Member States already have regulations and policy frameworks in place to support the empowerment of smaller and non-commercial market actors in the energy market. But this might become a disadvantage as Holstenkamp sketches in Chap. 6 with regard to the German situation where the turns and twists in energy policy leave RECs in a struggle to find new business models that work.

Second, the realisation of the CEP depends on economic and energy market developments. The development of REC business models depends on public regulations and market structures. On the basis of RED II, the participation of private undertakings in RECs is guaranteed as long as it does not constitute their primary commercial or professional activity. Furthermore, there is the criterion of 'proximity', which limits the number private undertakings in RECs. It cannot just be any private undertaking from anywhere. However, commercial partners might be involved not for profit reasons but for 'green washing', corporate social responsibility or other reasons for access to renewable energy markets.

Member States are to ensure that RECs can participate in available support schemes on an equal footing with large participants. They are to take into account any specific national conditions when designing support schemes, without prejudice to Articles 107 and 108 TFEU. This can give rise to tension due to grid connection licences based on a lack of grid capacity. For instance, an ICT company which plans to construct a data centre buys or organises a large part of the renewable energy in a region (Koronen et al., 2020) and gets priority based on regional economic considerations.

There is a difference between local energy projects and community energy projects. Although both modes of generating decentralised energy share a common focus on area- or place-based systems of provision (Devine-Wright, 2019) there are key differences. As discussed by Holstenkamp (in Chap. 6) and in the case of the energy cluster in Poland (see Chap. 3 by Ruggiero cs.) there is a risk of neoliberal ideas that frame RECs as non-efficient market parties. In the UK, policy on decentralized energy has shifted from community energy to local energy, which signals reduced support for grassroots, citizen-led action in favor of institutional partnerships and private sector company-led investment (Devine-Wright, 2019). Here, the risk is that projects are labelled as RECs but do not have the characteristics of a real civil society-based renewable energy community with actual citizen ownership, democratic control, autonomy and non-commercial goals.

Third, the realisation of the CEP depends on technological development. As renewable energy technology continues to develop, many energy communities are exploring the possibilities of storage, demand-response, grid balancing, bundling collectively generated energy into virtual power plants and microgrids (ECCO, 2021), among other things. These technological developments in combination with the CEP push for professionalisation of the community energy sector (see Holstenkamp in Chap. 6). However, the question can be raised whether RECs are willing—even with intermediary support (see Chap. 7 by Warbroek)—to professionalize?

Regulation will not develop as quickly as innovation. It typically follows or is developed in response to innovation taking hold, also in European energy markets (Hoppe et al., 2018). This forms a potential problem if RECs want to develop and implement innovative concepts, technologies and business models. For example, the Dutch tax system has been taxing battery storage systems twice, once when storing energy, and second when feeding it back into the electricity grid, which was limiting the integration of storage technologies. This will be solved in the Netherlands under the new Energy Act implementing the EU regulation with the CEP (Energy act, 2020), which targets energy systems to become more embedded and integrated systems. This also pays attention to other dimensions like heat and power (EU Strategy for Energy System Integration, 2020). However, there is currently a lack of EU rules on emerging energy market developments like storage and hydrogen. In general, the societal cost of grid capacity and grid connection makes the upscaling of RECs difficult. Another example is that although there are serious grid capacity problems there are still national governments that use economic incentives that promote feeding electricity into the grid instead of encouraging direct use first (i.e., pro-suming). In some cases this leads to problematic behavior because feeding electricity into the grid is compensated by a higher subsidy than using it. Some RECs might want to encourage a closed distribution system with limited feedback to the grid, or even closed systems or microgrids (autarkic systems) but this is often not considered rewarding.

11.7 Suggestions for Future Research

There are many differences between existing RECs. They come in all kinds of different shapes and sizes (Horstink et al., 2021) and they can engage in any energy-related activity, although the majority of RECs remain engaged in electricity generation—mostly solar and wind energy. In line with the four observations that are key to this volume, we shortly make some suggestions for future research.

First, the observation that the instrumental function of REC involvement in the energy transition has increased in importance. We concluded that there is a potential tension between the original idealistic motives of saving energy and producing renewable energy, and becoming serious players in energy markets. This raises the research question whether the tendency towards an instrumental perspective of community energy can influence the value and principles of their local, social, democratic and cooperative identities due to the call for scaling, professionalisation and new business models of RECs as consequences of the new market requirements.

Second, the observations about differences between front-running and late-starting countries require more research into the factors that explain the differences between EU countries. This should address both differences in transposing the CEP into national legislation and policy support mechanisms as well in other exploratory factors such as attitudes toward the cooperative model and the cultures of local energy activism.

The third observation concerned the trend of broadening the goals and activities of renewable energy communities over the years. This deals with the provision of these multiple energy services—more specifically if they become activities that are more commercial—and the consequences this might have with regard to the identity of RECs. This needs more academic attention.

The fourth observation concerned a mismatch observed between academic attention and the significance of the community energy movement. The community energy movement is treated in this body of literature as a social phenomenon with a strong focus on 'self-determination' of citizens as final customers. However, the energy transition is not one-dimensional but also entails other issues that are important and also relate to sustainable transitions, like the pathway to climate neutrality, decentralised and sustainable energy generation, the circular economy and the digital society. This requires a more integrated system perspective (Caramizaru & Uihlein, 2020) on the role of RECs within socio-technical systems. Socio-technical systems (Verbong & Geels, 2007) have already attracted had lots of attention in the sustainable transitions literature focusing on RECs as a social innovation and a developing niche (see Chap. 7 by Warbroek; Hoppe et al., 2015; Dóci et al., 2015). A different angle would be the acceptance of the necessary social and technical innovations. Social acceptance research tends to focus on public acceptance of support for specific energy generation projects (Wüstenhagen et al., 2007). More specific research is also needed on the acceptance of specific modes of decentralised energy provision and the relation between energy communities and the acceptance of energy saving practices, energy efficient and smart energy technologies, in particular in relation to demand control measures and flexibility (Coenen & Hoppe, 2018).

References

Aidukaiu, J. (2013). *Community Mobilizations Around Housing and Local Environment: Insights into the Case of Vilnius*. https://doi.org/10.15388/SocMintVei.2013.1.1846

Bauwens, T. (2016). Explaining the Diversity of Motivations Behind Community Renewable Energy. *Energy Policy, 93*(C), 278–290.

Bauwens, T. (2017). Polycentric Governance Approaches for a Low-Carbon Transition: The Roles of Community-Based Energy Initiatives in Enhancing the Resilience of Future Energy Systems. In N. Labanca (Ed.), *Complex Systems and Social Practices in Energy Transitions. Green Energy and Technology*. Springer. https://doi.org/10.1007/978-3-319-33753-1_6

Bauwens, T., Gotchev, B., & Holstenkamp, L. (2016). What Drives the Development of Community Energy in Europe? The Case of Wind Power Cooperatives. *Energy Research & Social Science, 13*, 136–147. https://doi.org/10.1016/j.erss.2015.12.016

Bocken, N., Jonca, A., Södergren, K., & Palm, J. (2020). Emergence of Carsharing Business Models and Sustainability Impacts in Swedish Cities. *Sustainability, 12*(4), 1594.

Caramizaru, E., & Uihlein, A. (2020). *Energy Communities: An Overview of Energy and Social Innovation*. EUR 30083 EN, Publications Office of the European Union, Luxembourg, ISBN 978-92-76-10713-2 (online), https://doi.org/10.2760/180576 (online), JRC119433.

Coenen, F. H. J. M., & Hoppe, T. (2018). D3.4 – Effectiveness Report 2: REScoop Plus Project Grant Agreement number: 696084.

Devine-Wright, P. (2019). Community Versus Local Energy in a Context of Climate Emergency. *Nature Energy, 4*, 894–896. https://doi.org/10.1038/s41560-019-0459-2

Devine-Wright, P., Batel, S., Aas, O., Sovacool, B., LaBelle, M. C., & Ruud, A. (2017). A Conceptual Framework for Understanding the Social Acceptance of Energy Infrastructure: Insights from Energy Storage. *Energy Policy, 107*, 27–31. https://doi.org/10.1016/j.enpol.2017.04.020

Dóci, G., Vasileiadou, E., & Petersen, A. C. (2015). Exploring the Transition Potential of Renewable Energy Communities. *Futures, 66*, 85–95.

ECCO. (2021). *Energy Community COoperations Project*. https://www.nweurope.eu/projects/project-search/ecco-creating-new-local-energy-community-co-operatives/

Energy act. (2020). *Contouren van de Energiewet (algemene toelichting)*. Technical Report, Ministerie van Economische Zaken en Klimaat, Den Haag, 7 2020.

EU Strategy for Energy System Integration. (2020). Communication from the Commission, Powering a Climate-Neutral Economy: An EU Strategy for Energy System Integration COM/2020/299 Final.

European Court of Auditors. (2019). *Wind and Solar Power for Electricity Generation: Significant Action Needed if EU Targets to Be Met* (Pursuant to Article 287(4), Second Subparagraph, TFEU).

Hess, D. J. (2018). Energy Democracy And Social Movements: A Multi-Coalition Perspective on the Politics of Sustainability Transitions. *Energy Research & Social Science, 40*, 177–189. https://doi.org/10.1016/j.erss.2018.01.003

Hewitt, R. J., Bradley, N., Baggio Compagnucci, A., Barlagne, C., Ceglarz, A., Cremades, R., & Slee, B. (2019). Social Innovation in Community Energy in Europe: A Review of the Evidence. *Frontiers in Energy Research, 7*(31). https://doi.org/10.3389/fenrg.2019.00031

Hoppe, T., Butenko, A., & Heldeweg, M. (2018). Innovation in the European Energy Sector and Regulatory Responses to It: Guest Editorial Note. *Sustainability, 10*(2), 416.

Hoppe, T., Graf, A., Warbroek, B., Lammers, I., & Lepping, I. (2015). Local Governments Supporting Local Energy Initiatives: Lessons from the Best Practices of Saerbeck (Germany) and Lochem (The Netherlands). *Sustainability, 7*(2), 1900–1931.

Horstink, L., Wittmayer, J. M., & Ng, K. (2021). Pluralising the European Energy Landscape: Collective Renewable Energy Prosumers and the EU's Clean Energy Vision. *Energy Policy, 153*, 112262.

Hufen, J. A. M., & Koppenjan, J. F. M. (2015). Local Renewable Energy Cooperatives: Revolution in Disguise? *Energy. Sustainability and Society, 5*(1), 18. https://doi.org/10.1186/s13705-015-0046-8

Internal Electricity Market Directive (IEMD). (2019). Directive (EU) 2019/944 of the European Parliament and of the Council of 5 June 2019 on Rules for the Internal Market for Electricity and Amending Directive 2012/27/EU.

Internal Electricity Market Regulation (IEMR). (2019). Regulation (EU) 2019/943 of the European Parliament and of the Council of 5 June 2019 on the Internal Market for Electricity.

Jetten, J., Mols, F., & Selvanathan, H. P. (2020). How Economic Inequality Fuels the Rise and Persistence of the Yellow Vest Movement. *International Review of Social Psychology, 33*(1), 2. https://doi.org/10.5334/irsp.356

Koronen, C., Åhman, M., & Nilsson, L. J. (2020). Data Centres in Future European Energy Systems – Energy Efficiency, Integration and Policy. *Energy Efficiency, 13*, 129–144. https://doi.org/10.1007/s12053-019-09833-8

Lennon, B., Dunphy, N. P., & Sanvicente, E. (2019). Community Acceptability and the Energy Transition: A Citizens' Perspective. *Energy, Sustainability and Society, 9*(35). https://doi.org/10.1186/s13705-019-0218-z

Lowes, R., Woodman, B., & Speirs, J. (2020). Heating in Great Britain: An Incumbent Discourse Coalition Resists an Electrifying Future. *Environmental Innovation and Societal Transitions, 37*, 1–17. https://doi.org/10.1016/j.eist.2020.07.007

Magnani, N., & Osti, G. (2016). Does Civil Society Matter? Challenges and Strategies of Grassroots Initiatives in Italy's Energy Transition. *Energy Research & Social Science, 13*, 148–157. https://doi.org/10.1016/j.erss.2015.12.012

Paravantis, J. A., & Kontoulis, N. (2020). *Energy Security and Renewable Energy: A Geopolitical Perspective.* IntechOpen. https://doi.org/10.5772/intechopen.91848

Proka, A., Hisschemöller, M., & Loorbach, D. (2018). Transition Without Conflict? Renewable Energy Initiatives in the Dutch Energy Transition. *Sustainability, 10*(6), 1721.

Radtke, J. (2014). A Closer Look Inside Collaborative Action: Civic Engagement and Participation in Community Energy Initiatives. *People, Place and Policy, 8*(3), 235–248. https://doi.org/10.3351/ppp.0008.0003.0008

Renewable Energy Directive (RED II) Directive (EU) 2018/2001 (Recast) on the Promotion of the Use of Energy from Renewable Sources.

Ruggiero, S., Busch, H., Hansen, T., & Isakovic, A. (2021). Context and Agency in Urban Community Energy Initiatives: An Analysis of Six Case Studies from the Baltic Sea Region. *Energy Policy, 148*, 111956. https://doi.org/10.1016/j.enpol.2020.1119560.3351/ppp.0008.0003.0008

Schwabe, J. (2020). The Evolution of Cooperative Electric Carsharing in Germany and the Role of Intermediaries. *Environmental Innovation and Societal Transitions, 37*, 108–119. https://doi.org/10.1016/j.eist.2020.08.007

Schwencke, A. M. (2021). Lokale Energie Monitor 2020. *The Hague*, 1–129.

Smith, A., Hargreaves, T., Hielscher, S., Martiskainen, M., & Seyfang, G. (2016). Making the Most of Community Energies: Three Perspectives on Grassroots Innovation. *Environment and Planning A: Economy and Space, 48*(2), 407–432. https://doi.org/10.1177/0308518X15597908

Som Energia. (2021). https://www.somenergia.coop/es/produccion/. Accessed 21 Apr 2021.

Van der Schoor, T., & Scholtens, L. J. R. B. (2021). *Scientific Approaches of Community Energy, a Literature Review*. Centre for Energy Economics Research (CEER).

Van Veelen, B., & van der Horst, D. (2018). What Is Energy Democracy? Connecting Social Science Energy Research and Political Theory. *Energy Research & Social Science, 46*, 19–28.

Vansintjan, D. (2015). The Energy Transition to Energy Democracy; Power to the People; Final Results Oriented Report of the REScoop 20-20-20 Intelligent Energy Europe project (D. Wrikker, Ed.). REScoop 20-20-20.

Verbong, G., & Geels, F. (2007). The Ongoing Energy Transition: Lessons from a Socio-Technical, Multi-Level Analysis of the Dutch Electricity System (1960–2004). *Energy Policy, 35*(2), 1025–1037.

Walker, E. T., & McCarthy, J. D. (2010). Legitimacy, Strategy, and Resources in the Survival of Community-Based Organizations. *Social Problems, 57*(3), 315–340. https://doi.org/10.1525/sp.2010.57.3.315

Walker, G., & Devine-Wright, P. (2008). Community Renewable Energy: What Should It Mean. *Energy Policy, 36*(2), 497–500.

Warbroek, B., & Hoppe, T. (2017). Modes of Governing and Policy of Local and Regional Governments Supporting Local Low-Carbon Energy Initiatives; Exploring the Cases of the Dutch Regions of Overijssel and Fryslân. *Sustainability* (Switzerland), *9*(1), 75. https://doi.org/10.3390/su9010075.

Westwell, E., & Bunting, J. (2020). The Regenerative Culture of Extinction Rebellion: Self-Care, People Care, Planet Care. *Environmental Politics, 29*(3), 546–551. https://doi.org/10.1080/09644016.2020.1747136

Wierling, A., Schwanitz, V., Zeiß, J., Bout, C., Candelise, C., Gilcrease, W., & Gregg, J. (2018). Statistical Evidence on the Role of Energy Cooperatives for the Energy Transition in European Countries. *Sustainability, 10*(9), 3339.

Wittmayer, J. M., de Geus, T., Pel, B., Avelino, F., Hielscher, S., Hoppe, T., Mühlemeier, S., Stasik, A., Oxenaar, S., Rogge, K. S., Visser, V., Marín-González, E., Ooms, M., Buitelaar, S., Foulds, C., Petrick, K., Klarwein, S., Krupnik, S., de Vries, G., & Härtwig, A. (2020). Beyond Instrumentalism: Broadening the Understanding of Social Innovation in Socio-Technical

Energy Systems. *Energy Research & Social Science, 70*, article 101689. https://doi.org/10.1016/j.erss.2020.101689

World Bank. (2017). *State of Electricity Access Report 2017 (Vol. 2): Full Report (English)*. World Bank Group. http://documents.worldbank.org/curated/en/364571494517675149/full-report

Wüstenhagen, R., Wolsink, M., & Bürer, M. J. (2007). Social Acceptance of Renewable Energy Innovation: An Introduction to the Concept. *Energy Policy, 35*(5), 2683–2691. https://doi.org/10.1016/j.enpol.2006.12.001

Yadoo, A., & Cruickshank, H. (2010). The Value of Cooperatives in Rural Electrification. *Energy Policy, 38*(6), 2941–2947. https://doi.org/10.1016/j.enpol.2010.01.031

Yildiz, Ö., Rommel, J., Debor, S., Holstenkamp, L., Mey, F., Müller, J. R., ... & Rognli, J. (2015). Renewable Energy Cooperatives as Gatekeepers or Facilitators? Recent Developments in Germany and a Multidisciplinary Research Agenda. *Energy Research & Social Science, 6*, 59–73.

Index[1]

A

Access to energy, 208–210, 212, 220
Asset-Based Community Development (ABCD), 162, 163

B

Baltic Sea Region, 17, 49–69
Baltic States, 16, 258, 266
Barriers, 6, 16, 45, 46, 51, 60–64, 68, 69, 77, 80, 89, 91, 92, 114, 154, 156, 158, 160, 161, 166, 169, 210, 214, 215, 222, 251, 268, 273, 276
Belgium, 5, 17, 46, 190
Bioenergy, 133
Biofuel, 84
Bottom-up, 47, 101, 112, 114, 161, 171, 228n1, 236, 241, 264
Buildings, 61, 68, 69, 79, 80, 85–87, 89, 90, 102, 109n2, 132, 155, 158, 160, 161, 165, 169, 201, 202, 231–233, 236, 238, 240–245, 247–250, 268, 270, 271
Business incubator, 161

C

Cases, 15, 25, 53, 55, 58, 59, 63, 67, 68, 76, 78, 81, 82, 87–92, 101–104, 114, 120, 126, 127, 129, 130, 134, 144, 170, 179, 190, 231, 238, 240, 241, 244, 259, 261, 264, 267–269, 271, 272, 276

[1] Note: Page numbers followed by 'n' refer to notes.

© The Author(s), under exclusive license to Springer Nature Switzerland AG 2021

F. H. J. M. Coenen, T. Hoppe (eds.), *Renewable Energy Communities and the Low Carbon Energy Transition in Europe*, https://doi.org/10.1007/978-3-030-84440-0

Case studies, 16, 51, 55, 63–65, 68, 76, 81–86, 165, 190, 198, 201, 228, 239, 240, 249, 272
Citizen Energy Community (CEC), 24, 25, 27–38, 40–44, 49, 108–111, 116, 140, 142, 156, 180, 180n1, 181, 202, 217, 218, 258, 259
Civil society, 11, 60, 69, 75, 77, 78, 84, 86, 89, 210, 266, 267, 273, 277
Clean energy, 12, 23, 75–93, 206, 212, 213, 258, 260
Clean Energy For All Europeans, 2, 24, 207, 213, 219, 245, 257
Clean Energy Package (CEP), 14, 17, 23–47, 98, 120, 140–143, 154–156, 169, 181, 182, 209, 228, 257, 258, 261, 269, 270, 273–279
Climate, 1, 6, 7, 24, 31, 42, 160, 202, 205, 206, 228, 246, 257–262, 272, 279
Climate action, 13, 206, 215
Climate agreement, 6, 156
Co-housing, 79, 227, 228, 228n1, 231–251
Commission, 24, 25, 25n6, 27n11, 28, 33n24, 36n28, 47, 50, 275
Common rules for the internal market for electricity, 24n2
Communities, 2, 23, 52, 76, 97, 119, 153, 180, 206, 227–251, 258
Community based, 75, 161
Community development, 77, 104
Community energy (CE), 2, 24, 49–69, 75, 97–117, 119–144, 258
Community energy drivers, 63–64
Community Renewable Energy, 130–131
Competition, 25, 26, 28, 34, 46, 194, 261
Consumers, 7, 14, 26, 30, 32, 33, 38, 39, 42, 53, 60, 99, 109, 179, 186, 206, 207, 210, 213, 215, 217–221, 223, 238, 250, 258, 271, 274
Cooperative, 3, 23, 57, 78, 98, 120, 155, 184, 241, 260
Co-production, 202
Crowdfunding, 44, 106, 114

D

Denmark, 4–6, 9, 17, 23, 29n16, 50–52, 57–59, 108, 119, 190, 271, 272
Development, 7, 24, 51, 77, 97, 120, 153, 181, 206, 238, 258
Directive, 6, 8, 11, 13, 15, 24, 30, 31, 35, 39, 49, 50, 68, 69, 117, 140, 153, 169, 228, 240, 250, 262, 266, 274, 275

E

Economics, 1, 2, 6, 8, 11–13, 30, 31, 36–38, 57, 63, 66–68, 77, 82, 85, 91, 101, 103–104, 110, 113, 130, 132, 133, 143, 162, 181, 183, 187, 206, 208, 210, 214, 216, 218, 240, 245, 262, 265, 268, 273, 275–278
Efficiency, 2, 9, 11–14, 26, 31, 39, 61, 62, 104–106, 143, 179,

181, 182, 191, 194–198, 213, 215, 218, 222, 243, 245, 246
Electricity act, 156
Electricity directive, 2, 11, 49, 68, 108, 111
Endogenous development, 162, 165
Energiewende, 119
Energy, 1, 24, 49, 53, 153, 179, 205, 227–251, 258
Energy Act, 156, 274, 278
Energy agency, 58, 115
Energy agreement, 155
Energy community, 2, 4, 7, 23–47, 68, 75, 98–100, 113, 115, 116, 120, 127–143, 153, 179–202, 205–223, 228, 238, 240, 245–247, 250–251, 258, 260, 262, 265, 267, 269, 270, 274, 275, 277, 280
Energy cooperative, 4, 9, 11, 14, 76, 78, 82, 85, 89, 98, 114, 115, 128, 130, 132, 133, 139, 143, 144, 155, 156, 184, 187, 189, 191, 194, 198, 199, 201, 241, 267, 270, 274
Energy provider, 2
Energy Service Company (ESCOs), 110, 113, 114, 116, 117, 159
Energy tax, 155
Energy transition, 4, 6, 7, 10–15, 17, 23, 25, 26, 41–42, 41n39, 47, 75–92, 100, 101, 114, 115, 119, 121, 131, 156–158, 160, 179, 205–218, 220–223, 228, 239, 241, 246, 250, 258, 260–262, 265, 278, 279
Energy Union, 13, 25, 215
Energy Union Strategy, 25
England, 5
Environment, 52, 141, 157, 180, 181, 190, 196, 200, 237, 240, 246, 251
Estonia, 17, 51, 55, 57, 59–61, 63, 66–68, 258, 266
EU Clean Energy, 238
EU Clean Energy for All Europeans Package, 245
EU Commission, 34, 46, 46n46, 219
EU directive, 3, 6, 10, 14, 15, 69, 82, 92, 98, 100, 108–110, 116, 215, 219, 274
EU Green Deal, 222
EU Guidelines, 275
EU Member States, 7, 11, 43, 92, 121, 142, 209, 214, 215, 258, 259, 266–269
EU Renewables Directive, 2, 10, 13
European Commission, 24, 60, 68, 207, 209, 214, 221
European Energy Market (EEM), 277
European Green Deal, 258, 270
EU Strategy for Energy System Integration, 258, 270, 278

F

Finance, 47, 68, 91, 92, 106, 110, 112, 113, 161, 199, 202, 239
France, 5, 17, 92, 190, 261

G

Gender, 182, 205–223, 261
German, 45, 119–121, 126–143, 274–276
German Energiewende, 119

Global South, 206, 208, 209, 222, 223
Governance, 13, 26, 28, 34, 36–38, 42, 43, 77, 84, 101, 131, 141, 142, 156, 159, 211–213, 263
Government, 8, 9, 13, 44, 45, 52, 53, 60, 77, 121, 129, 132, 140, 142, 143, 153–155, 158, 159, 166, 169–171, 180, 181, 183, 184, 186, 198–199, 202, 206, 213–215, 219, 240, 241, 260, 263, 268, 278
Grassroots, 79, 101, 264, 277
Greece, 16, 17, 75–92, 218, 258, 264, 266–268, 272–275

H

Home energy savings, 179–202

I

Intermediary Organizations, 158, 165
Internal Electricity Market Directive (IEMD), 24, 26–32, 34, 35, 38, 40, 41, 43, 45, 202, 258, 259

L

Latvia, 17, 51, 55, 57, 59, 61–62, 68, 258, 266, 268
Legislations, 2, 11, 15, 30n18, 34, 37, 42, 44–47, 61, 68, 69, 78, 111, 153, 155, 156, 202, 207, 213–215, 217, 218, 222, 258, 259, 262, 266, 271, 274, 276, 279
Lithuania, 17, 51, 55, 57, 59–63, 66–68, 258, 266
Living, 68, 110, 134, 214, 215, 237, 249
Local authority, 4, 11, 33, 34, 36, 42, 50, 62, 75, 76, 78, 86, 92, 101, 110, 126, 129, 131, 134, 264, 265

M

Management, 34, 110, 161, 166, 227, 231, 238, 242, 247, 249, 251
Member State, 7, 13, 15, 24, 26, 27n11, 29, 32–35, 37, 39–42, 44–47, 69, 153, 156, 213–215, 218, 258, 274–277

N

National Energy Climate Plans (NECPs), 13, 42, 140, 209, 214–220, 222
National Transposition of New EU Directives, 41n39
Netherlands, 3, 5, 6, 9, 16, 17, 23, 50–52, 92, 154–156, 190, 194, 198, 201, 217, 221, 227–251, 258, 266, 268, 271, 272, 274, 275
NGOs, 88, 159, 264

P

Paris Agreement, 12, 206
Photovoltaics, 99
Poland, 5, 16, 17, 51, 55, 57, 59, 61–62, 68, 258, 264, 266, 268, 273, 276, 277

Policy, 2, 23, 51, 53, 63, 77, 99, 120, 156, 180, 205–223, 231, 258
Poverty, 11, 26, 30, 31, 181, 187, 206, 209, 213–218, 220–223, 261
Public, 6, 9, 12, 29–31, 35, 75, 78, 82, 88, 90, 101, 102, 107, 112, 114, 120, 129, 132, 134, 138, 162, 167, 180, 181, 184, 186, 199, 240, 242, 261, 263, 270, 276, 279
Public policy, 183, 184, 212
PV-generation, 246
PVs, 84, 85

R

REC movement, 171
RED II Directive, 14
Regulation, 15, 42, 47, 66–68, 78, 91, 101, 110–112, 114, 116, 120, 129, 140, 142, 156, 231, 240, 265, 268, 269, 271, 274, 276–278
Renewable, 52, 59
Renewable energy, 49–52, 55, 60–62, 67–69
Renewable energy communities (RECs), 1–6, 8–16, 24–31, 33–39, 41n39, 42–44, 49, 50, 55, 62, 98, 100, 108–115, 140, 142, 153–171, 180, 215, 217, 218, 222, 228, 245, 257–259, 262–268, 270–274, 276–279
Renewable energy supplying cooperatives (REScoops), 3, 11, 76, 78, 80, 81, 83, 85–91, 186, 188–190, 195–197, 201, 245, 261, 263, 264

S

Social inclusion, 210, 218, 220, 247
Strategic Niche Management (SNM), 17, 76, 79–80, 82, 86, 88, 90, 160, 161
Sustainability, 77, 87, 88, 91, 189, 240, 242–244, 247, 248, 250

T

Technology, 23, 26, 55, 66, 77, 78, 82, 89, 99, 103, 107, 110, 113, 119–144, 155, 167, 194, 207, 219, 236, 240, 241, 243, 244, 248, 249, 251, 261, 277, 278, 280
Theory of Planned Behavior, 182

U

United Kingdom (UK), 5, 9, 29n16, 50, 51, 140, 220, 271, 272

W

Windenergie, 138

The manufacturer's authorised representative in the EU is Springer Nature Customer Service Centre GmbH, Europaplatz 3, 69115 Heidelberg, Germany. If you have any concerns regarding our products, please contact ProductSafety@springernature.com

Printed and bound by CPI Group (UK) Ltd, Croydon, CR0 4YY
30/09/2026
02225492-0004